D0909778

ANIMALS AS NAVIGATORS

BY THE SAME AUTHOR

The Principles of Navigation

ANIMALS AS NAVIGATORS

E. W. Anderson

O.B.E. D.F.C. A.F.C. M.SC. F.R.MET.S. Hon. M.R.I.N.

Royal Institute of Navigation
Gold Medallist

VAN NOSTRAND REINHOLD COMPANY
NEW YORK CINCINNATI TORONTO LONDON MELBOURNE

To Betty

Library of Congress Catalog Card Number 82–21898
ISBN 0–442–20882–0

Printed in Great Britain

Published by Van Nostrand Reinhold Company Inc.
135 West 50th Street
New York, New York 10020

16 15 14 13 12 11 10 9 8 7 6 5 4 3 2 1

Library of Congress Cataloging in Publication Data

Anderson, E. W (Edward William), 1908–
Animals as navigators.

Bibliography: p.
Includes index.
1. Animal navigation. I. Title.
QL782.A53 1983 591.1 82–21898
ISBN 0–442–20882–0

CONTENTS

List of Diagrams and Sketches

Figs. 28 and 35 have been drawn by the author. Figs. 1, 6, 12, 13, 15, 18–22, 24, 26, 27, 29, 33, 36–43 have been drawn by Linda Page. Figs. 3, 4, 5, 7, 10, 11, 14, 16, 17, 23, 31 and 34, drawn by the author and lettered by Richard Reeves, are reproduced from *The Journal of Navigation* by courtesy of The Royal Institute of Navigation.

List of Photographs

Author's Preface

Nobody can read what has been written about animal navigation without a growing understanding of and love for the creatures with whom we share this planet.

It appears that previous studies on animal navigation have been based on traditional marine systems and are concerned principally with finding the way and steering. Navigational aids, such as radar fitted to naval vessels and military aircraft, are nonetheless necessary to get into position to strike at or to evade the enemy, in other words, for the positive and negative aspects of hunting. Today, hunting and avoiding up to the point of the final attack are accepted as an integral part of navigating.

Because the search for food is paramount, animals tend to find their way by using the systems they have evolved for hunting, rather than *vice versa*. Thus a seabird foraging in the wide oceans can be expected to home by methods quite different from those of the fox hunting locally by sight and smell. But it is surely wrong to assume that, because the latter tends to become lost if transported artificially into an area it does not know, it is any the less efficient as a navigator.

Certain fields of human navigation can suggest why animals behave in particular ways. The habits of skin-divers and of pilots in light aircraft can explain the figure-of-eight dance of the honey bee and why it is changed when the nectar area is close. Elementary wave theory shows that animals can use radio only in a very restricted fashion, and gives the reason for small creatures detecting ultra-violet rather than infra-red. The sonar designer's preference for ultrasonics emphasises the high aural directivity of rustling sounds; hence rabbits employ drummings for alarm signals. Moreover, navigation gives a simple explanation for the lop-sided ear-holes of toothed whales and the asymmetrical ear-flaps of owls.

In this book I have attempted to illustrate the interface between human and animal navigational skills. For instance, hunting for moving prey frequently involves a chase. Classical methods of interception, with the requirement to know the course and speed of the quarry, are obviously out of the question for animals but 'proportional navigation', developed for anti-aircraft missiles, solves the problem immediately and continuously, provided the hunter can measure changes of direction. This book contains an experiment to show the links between eyes and 'rotation sensors' but the need to steer requires more than this.

Again, zoologists have been aware that animal accelerometers and the so-called 'rotation sensors' (for it is only certain flies that can detect rotations) serve to prevent an animal falling over by defining the vertical, which is a general rather than a navigational application, necessary even when standing still. But the vertical is an essential element in steering, whatever the system used, which is why the needle in the simple hand-compass has to be pendulously mounted. Furthermore, use of the vertical will be shown to enable an animal to neglect the changing azimuth rates of the Sun, Moon and stars, which vary with the passage of time, and to ignore cross-winds when airborne over land.

Vertical sensors contribute in other ways. Accelerometers are found to act as 'rotation sensors' and allow an animal to travel in company or to stop at a chosen point, enabling a bird to land on a swaying branch. Even walking towards an object seen ahead involves the accelerometers, the 'rotation sensors' and the edges of the eyes. However, the vertical probably affects position-finding to a lesser degree, but it simplifies homing when using the Sun, and may make it possible for salmon to return to each river on more or less the same date. Indeed, there may be applications yet to be discovered. The Polynesians learnt to find the way by reference to the vertical and that alone, lying on their backs in their catamarans, as we discover from studies of 'primitive' peoples' navigation.

On the negative side, the animal vertical is found to have a short memory, which precludes the use of inertial navigation techniques, gyrocompassing methods and measurements of Coriolis. Accuracy is limited, which will be shown to explain why pigeons find difficulty in homing when the countryside is unfamiliar and they are released *less* than seventy miles away, and why rudimentary animals either move slowly or dart about. The constituents of timing accuracy are also discussed and related to errors in homing when a bird is away from its nest for many days.

In the past, so many developments in engineering have been

sparked off by the researches of animal specialists that a contribution in return from human navigation and control theory is long overdue. Fresh ideas stimulate, even if the initial effort is towards disproving them, and some of the techniques here suggested will be new even to human navigators. Nevertheless, this book started out simply as a review of animals as navigators. The basic facts were extracted from the writings of zoologists, who are noticeably scrupulous in attributing ideas to earlier researchers. Unfortunately, it has often been possible to give references only to these original sources, particularly when several later writers have developed ideas in parallel.

People in the field of animal behaviour will recognise how much dependence has been placed on past-masters such as G. H. Allen, Maurice Burton, A. Butenaudt, A. Dendy, L. R. Donaldson, V. B. Droscher, D. R. Griffin, A. D. Hasler, Gustav Kramer, H. Lissmann, K. E. Machin, L. J. and M. Milne, F. P. Mohres, E. P. & Mrs. Sauer, K. Schmidt-Koenig and Wolfgang and Roswitha Wiltschko. Nevertheless there must be a special word of admiration to K. von Frisch for his work on bees and to Dr. G. V. T. Matthews for his suggestions on how birds use the Sun.

The work was originally undertaken for the Duke of Edinburgh's lecture given at the Royal Institute of Navigation in 1981. Subsequently, the Director of the Institute, Michael Richey M.B.E., suggested the paper be developed into a book. I hope he will be pleased with the result. A great deal is owed to Guido Waldman, publisher and friend, for his numerous valuable suggestions. Finally my thanks to Mrs. G. B. Elwertowska for the checking of so many details.

E. W. ANDERSON

Introduction

The ant finds its way back to the nest using a small part of its computer which is one quarter the size of a pin's head but infinitely more flexible than the silicon chips clumsy humans employ. Fragile butterflies flutter thousands of miles to congregate together in certain trees. Our astonishment at what quite primitive creatures can achieve must not blind us to the sober fact that there is nothing occult about finding the way, though it has suited some people to pretend there is.

Navigation has had its high priests, like those of Egypt who were wont to prophesy the rise and fall of the sacred river Nile with great pomp and ceremony, much magic and mystery, and a very small pipeline direct to the river itself. Although animals are such marvellous navigators, we must search for the little pipelines. The comings and goings of creatures may be incomprehensible today but tomorrow there will be a down-to-earth solution.

Zoologists try to visualise how animals navigate and then set to work, with brilliant ingenuity, to prove or disprove their theories so far as they can. We shall follow their example and suggest methods by which living creatures produce such wonderful results, taking care that our suggestions do not run counter to the researches of these great men and women. It must be left to them, in the future, to prove or disprove what, in navigational terms, seems likely.

ANIMALS

It may be useful first to remind ourselves of the broad range of creatures which, unlike plants, need to move from place to place for their food and we shall look at them briefly through the eyes of a navigator. The first of them seem to have been very primitive entities populating the oceans where the temperatures were reasonably low and the complex chemicals from which they were built tended less to decompose. Some of these creatures developed the ability to withstand being stranded and eventually took to the land. Many such animals have survived in some form today, including those which remain as simple cells.

A particular problem is the relationship between surface area and weight. Given two similar creatures but one ten times as long, wide and deep as the other, the smaller will have one hundredth of the area of the larger but only one thousandth of the volume and the weight. Thus its weight will have been reduced ten times as much as its area.

Just as a speck of dust, with practically no weight for its area, will stick to the side of a glass, so a very small creature lives in a world of surface tensions rather than gravity. It will stick to the underside of anything or float through the air attached to a particle of moisture. We need not examine the navigational problems of such microbes for, in general, their motions will be determined by their surroundings and not by themselves.

From such simple beginnings, marine arthropods appeared, with jointed legs and hard outsides tending to restrict growth. Indeed, a few quite complex specimens have remained no bigger than a large single cell. From them came the crustaceans, such as shrimps and crabs, which colonised the sea, and the insects that took to the land, some developing wings, like the beautiful butterflies which travel remarkable distances, particularly on the winds.

Also from the primitive marine creatures emerged the first vertebrates, fish with backbones and streamlined bodies, capable of foraging fast and far, and with skins that allowed them to grow. From them, all the higher animals developed. Some crawled onto the land at a later stage of their lives and hence the newt and the frog. From such amphibians came the reptiles, still producing their young in water but more safely within eggs. These two types of animal generally had legs but, as they protruded sideways like the fins from which they had evolved, they were better for swimming than for travelling over the ground.

Reptiles took to the air to become birds, converting scales into feathers that retain heat very well and allow them to control their temperatures, which reptiles can only partly achieve. Also there appeared warm-blooded furry mammals whose females carried young in water within their bodies. These animals had legs underneath, giving them a greater speed potential over the ground and some could cover long distances.

The relationship between surface area and weight also affects warm-blooded creatures. The smaller, the less will be the weight compared to the area and so the less relatively will be the energy to burn up and keep the skin warm. Thus there is a limit to the smallness of birds and mammals. Also, the heavier the flying animal, the smaller its wing area in proportion. This gives an explanation for bats being small and, though there are other factors, it is true that the heaviest birds do not fly.

Many creatures have returned to the habitat of their distant ancestors. Some winged insects have gone back to the land and a few birds no longer get airborne, penguins living largely in the sea. The huge whales and their smaller cousins the dolphins have taken to the oceans — but still breathe air — and their tails, attached like flippers to what were hind legs, lie horizontally, not vertically as in fishes. All this wonderful variety of creatures now inhabits our planet but it is obvious that the great navigators will be found among the flying insects, fish, swimming reptiles, birds and to some extent mammals, whales in particular travelling long distances as they do not have to negotiate unfavourable terrain.

In Fig. 1, humans are included although many people do not class

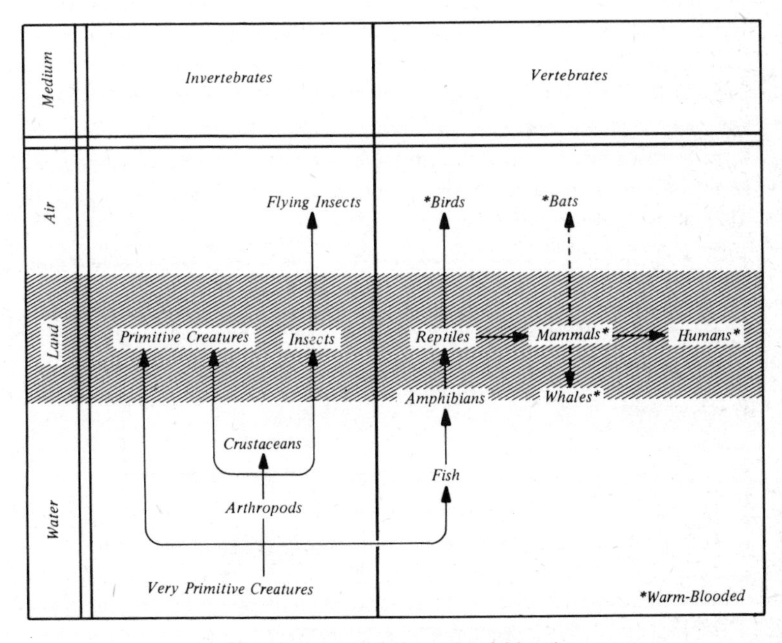

1. Development of the animal kingdom

them as animals. However, we shall often refer to the so-called 'primitive' peoples who have learnt to find their ways without any of the tools used by mariners or airmen today. For they or their descendants may be able to communicate with us and explain how they used to manage, and this can give valuable indications of the methods that may be employed by animals.

NAVIGATORS

We now turn from animals to the second word in the title of this book which obviously is linked to navigation, a term that originally meant 'ship-driving'. Today it embraces the journeyings of space-craft, nuclear submarines and supersonic aircraft not to mention ants and antelopes, the retention of the term being a tribute to those who founded the science at sea. Nevertheless any modern definition must apply to vessels sailed by mariners.

A ship at anchor is not navigating for it has to be on the move. Yet spinning round and round like a top is not sufficient, the *motion* has to be from one *place* to another. Nevertheless, drifting with the tide is hardly navigating, for the motion from place to place needs to be under *control*. It must also be *purposeful*, for aimless sailing hither and thither will not fill the bill. The definition of navigation therefore emerges as *purposeful control of motion from place to place*.[1] This will apply to all craft and also to animals if we use each word in its correct sense. Thus it must form the basis on which our book is built, and therefore it appears in Fig. 2.

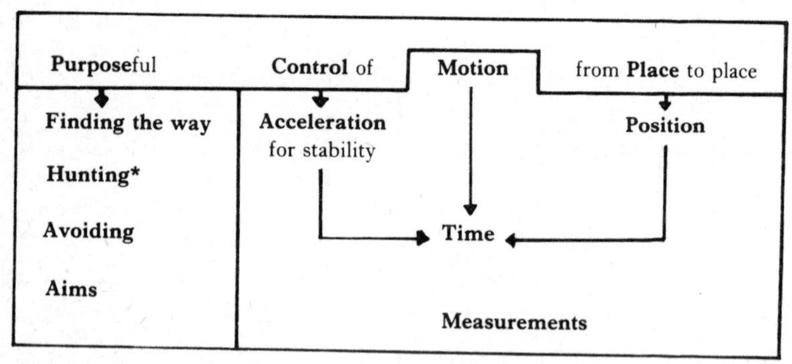

* includes finding a mate.

2. The pattern of animal navigation

Let us look first at purposes. It seems that only an animal whose journey has a purpose, can be classed as a navigator and this implies a certain level of intelligence. Indeed, in human beings, the ability to navigate seems likely to be a reasonable test of this quality. It follows that very primitive creatures, whose movements are controlled by outside forces or influences and have no say in which way to go, can hardly be classed as navigators. They have to be very primitive, for even an earthworm will make a choice. Put it in a tube with a T junction at the end and arrange that if it turns one way it will be given a slight shock and it will soon learn to choose the other.

Let us next analyse what are the main categories of purposes. A cargo vessel aims at reaching a port and a migrant swallow finds the way to Africa. Alternatively, the objective may be moving. A naval frigate searches for a submarine and a tiger hunts for prey and also at times for a mate. There is a general need at sea, on land and in the air to avoid collisions with other craft, and many animals spend much of their lives avoiding enemies. So the three purposes may be classified in animal terms as

(a) Finding the way,
(b) Hunting and
(c) Avoiding.

To achieve these purposes, it is necessary to control motion from place to place. *Motion* is the key but control is not only a matter of adjusting motion by *acceleration* or deceleration but also of maintaining motion unchanged. For the word 'control' implies restraint[2] of unwanted accelerations to ensure *stability* of motion. Place is, of course, *position* while *time* links motion to acceleration and, more importantly, to position. Indeed, the early Polynesians had no word for distance other than time taken to travel the journey. Thus we see that four elements have to be measured, namely:

(a) Acceleration for stability,
(b) Motion,
(c) Position and
(d) Time.

The three purposes and the four measurements appear in Fig. 2 and between them suggest the way in which our book ought to be arranged. We know animals carry stabilizers to keep themselves upright and, were this all, relegation of such sensors to the ends of books on navigation, or even their complete omission, would be logical. Yet stabilizers do much more than this. Man has to fit them to individual instruments and aids. Ships' compasses are mounted in gimbals, magnetic outputs in high-speed craft are also steadied by gyroscopes, and gyrocompasses are in fact stabilizers. The directional loop and the radar receiver depend on stabilized instruments and radar antennae are also independently levelled.

An animal does not carry separate balancers attached to eyes and ears but relies on centralised stabilizers linked to the individual senses by the brain. Thus stability is fundamental and has to be looked at first, not last. So essential is it that, for example, in the human baby the stabilizers are fully formed and of adult size four months *before* birth.[3]

We therefore begin with stabilizers and then turn to time and motion. These three will form the fundamentals of our study. The senses used by animals are applied in their simplest and most basic form to enable them to hunt or to avoid being caught. Finally, we shall discuss how all that has gone before can be used to find the way. Thus our book will be divided into three parts:

(I) Fundamentals,
(II) Hunting and Avoiding and
(III) Finding the way.

PART ONE

Fundamentals

I

Stabilizers and Time

SENSORS

Everything on the Earth is attracted by gravity, that powerful force which accelerates objects towards the centre of our planet. It affects all things equally and so it cannot be detected. All that can be sensed is the force that offsets it, the pressure on the soles of the feet or on the back when lying down and gazing skywards. That is why, within a module in outer space not subjected to any reaction by the vacuum around, everything feels weightless. On the Earth, gravity acts downwards and gives weight to objects, but we can only measure this weight by detecting the upward forces that counteract it.

It has to be either an animal so small it lives in a world of surface tension, or a very primitive creature of vague uncertain shape, not to have a right way up. Even the jellyfish, ninety-five per cent water and virtually no brain, working its way about by expanding and contracting its skirts, needs to know the way to the surface or to the bottom. The earthworm, almost as rudimentary, likewise requires reassurance it is burrowing downwards from a bird or upwards for dew.

Primitive jellies in arctic or antarctic waters may use the Earth's magnetic field, which is nearly vertical around the poles, to tell them which way is up. The water-boatman, that comical little creature which rows itself so vigorously across our ponds, carries an air bubble between its antennae. Turn it over and the bubble bends the antennae the wrong way and the little fellow hastily turns the right way up.

Thus it employs the surface of the water instead of an accelerometer, a procedure it shares with the mariner using a sea sextant. However, nearly all animals rely on detecting accelerations.

Accelerometer. A scallop stranded on the beach digs downwards into the sand, steered by a stony particle in a sac within its foot. The sac is full of liquid with hairs inside that bend to the weight of the particle. If the hairs at the bottom of the sac are affected, as in Fig. 3(a), the creature knows it is the right way up. If the mollusc tilts, as shown in Fig. 3(b), other hairs support the particle, a message travels to a sorting office and appropriate signals are sent to muscles to restore the body to upright, in this instance by a clockwise rotation.

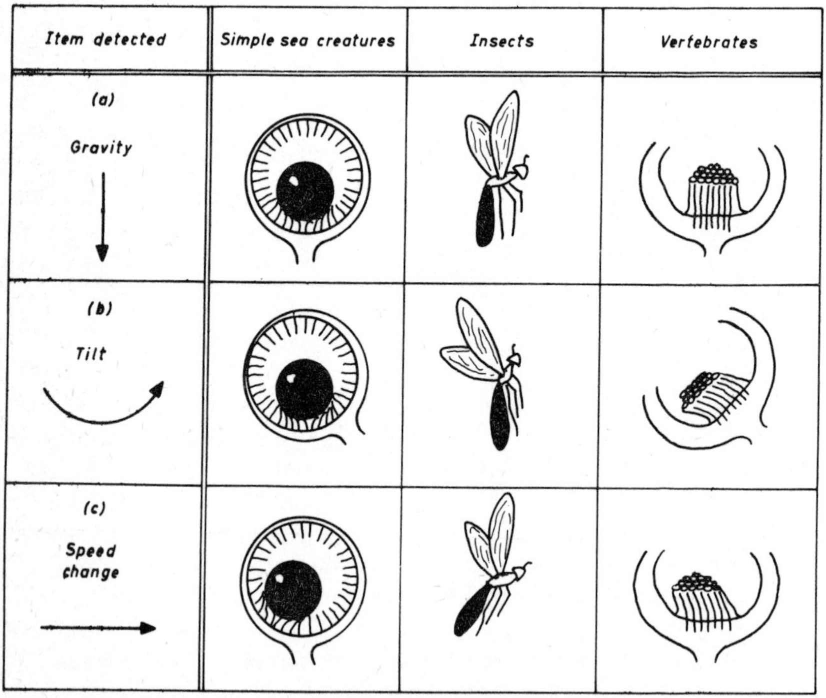

Item detected	Simple sea creatures	Insects	Vertebrates
(a) Gravity			
(b) Tilt			
(c) Speed change			

3. Accelerometers

Although we have illustrated a round object in a sac, the details of such sensors vary greatly. A crayfish, for example, has a sac open at the top so that it fills with seawater and grains of sand take the place of the particle. When the creature moults, it loses the skin of its sacs and the sand inside, and has to shovel more sand into the new pits by means of its claws. A moulting crayfish in an aquarium was given iron filings instead of sand and these were duly shovelled in. A magnet was placed above the tank and attracted the bits of metal upwards. The deluded crustacean, in order to keep these new objects pressing against the hairs at the bottom of its sacs, promptly turned upside down.

Suppose, as in Fig. 3(c), the mollusc were to start moving from left to right. The particle, being denser than the liquid, would lag behind and would press against the hairs in the sac not at the bottom but to one side as shown in the figure. Therefore the hairs would give the animal's brain the same message as if the sac had been tilted, and there will be no way to distinguish between the two. It is, of course, acceleration that is being registered, but we shall use the expression 'speed change' because gravity is also an acceleration and we need to distinguish between the two.

A change of speed may not matter to a scallop, for gravity is an acceleration of 32 feet per second every second and the creature moves so slowly the particle in the sac will hardly be shifted. Nor will it be a serious problem to an animal that hops about or darts from place to place, the change of speed at the start of the jump being cancelled out by the reversal of speed at the end. Provided the hops are only short and the creature stops at the end of each for a brief moment, it will be possible to ignore what the sac registers in between. This explains why so many tiny sea, land and air creatures dart about.

Although related to crustaceans, insects do not carry sacs. Instead, they employ sensors on their bodies in the form generally of short hairs outside the joints of their legs, round their waists and over their necks. If they fly, legs and abdomens hang down and gravity, tilt or speed change is detected as suggested by Fig. 3. When they walk, they can also sense the support given by the ground to their feet and by the way their legs react to equalise the pressures.

Once again, confusion between tilt and speed change will not matter if the insect crawls about slowly or darts from place to place. But consider a fly taking off. The forward speed change will cause abdomen and legs to hang behind as in Fig. 3(c) and, if deprived of the means to know better, the fly will imagine it is tilted upwards as in

Fig. 3(b). In correcting for this, it will nosedive straight into the ground.

The type of vertical sensor used by certain crustaceans and by all animals with backbones consists of a pile of chalky granules supported on a layer of jelly in a large complex sac filled with liquid, the position of the jelly being sensed by hairs projecting from below. This is shown in Fig. 3(a), and we may describe it as a 'sandwich-sensor'. If we tilt the sac, the sandwich will be warped sideways as in Fig. 3(b) and a speed change shown in Fig. 3(c) will distort it in the same fashion.

'Rotation' sensors. A hunter needing to sustain his motion or an animal that has to escape an enemy will have to distinguish between a tilt and a speed change, otherwise it may act the wrong way. Accordingly, certain crustaceans have developed sacs filled with liquid and with long hairs as shown in Fig. 4(a), the absence of a particle ensuring there is no reaction to gravity nor to speed change.

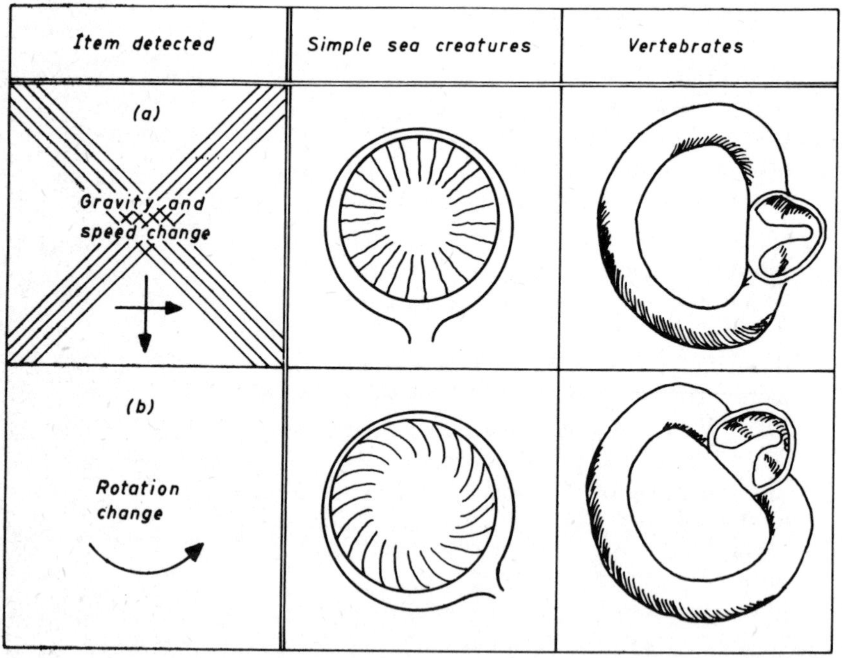

4. 'Rotation' sensors

However, when the sac tilts, as in Fig. 4(b), the liquid remains still, just as tea in a cup does not turn when the cup itself is rotated. Therefore the hairs are pulled round by the liquid and bent sideways.

If we put a crab on a turntable and start it up, the eyes on stalks turn at first, trying to keep things in view. However, when the rotation settles down the eyes centralise, which shows rotation is not being recorded. Increase speed and the eyes turn the same way. Decrease it and they turn the opposite way. So the so called 'rotation' sensors really detect changes of rotation: as the hairs are bent, they tend to set the liquid in the sac rotating in the same direction.

The octopus, a creature with a reputation for intelligence, though it has been known to devour its own tentacles, lives by hunting and is also pursued by moray eels and by whales. It has to be able to move quickly without getting confused or tilting itself unnecessarily. It carries 'rotation' sensors in the form of a bag filled with liquid but uses flaps of jelly with hairs inside to find out when it is starting to rotate against the static liquid. These flaps are arranged in sets of three, each at right angles to the others, and so the octopus can detect a change of rotation in any direction.

Vertebrates carry tubes known as canals, filled with liquid and in the form of a semicircle with the ends connected by a straight channel as shown in Fig. 4(a). Along this diameter and at right angles to the plane of the semicircle is a flap of jelly, with hairs protruding from below. If the canal starts to turn in one direction, as shown in Fig. 4(b), the liquid inside tends to stay still and pulls the flap the other way. Experiments prove the flap bends initially but returns slowly to the point of rest as it causes the liquid to swirl in the canal at the same rate as the canal is turning.

If rotation is then increased, the flap bends the same way as before, until the liquid is made to catch up with the faster turning. If the rotation is slowed or stopped, the swirling liquid comes up against the flap and bends it the opposite way. It is possible that, at the initial bending, the semicircular canal may register rotation but it very quickly becomes a sensor of change of rotation or angular acceleration. The same thing may happen with the simple hairs in the sac or with the flaps of the octopus.

Although these sensors basically measure change of rotation, they are generally referred to as 'rotation' sensors in books on animals. This is perhaps reasonable, for the brain of an animal can convert change of rotation into rotation by a process of summarising or integration. We shall therefore continue to use the term 'rotation' sensors, the inverted commas reminding us that they really detect rotation changes.

An insect, as usual the odd man out in terms of stabilizing sensors, has special ways to measure 'rotations'. As we have seen, there are no problems if it moves about very slowly nor if it jumps from place to place, speed change signals on take-off being cancelled out by those on landing. If it flies, it generally has four wings and, in the same way that the legs keep an insect level on the ground, so the wings may help to control flight. Besides, the wings of insects flap between upright and horizontal and have a general upwards tilt which keeps balance automatically, just as a sheet of paper rolled into a cone falls point downwards. Nevertheless, one suspects their systems are not particularly effective.

Dragonflies, which travel very fast, have heavy heads and it requires more energy to rotate them than other parts of their bodies. Tufts of bristle round their necks can therefore sense angular accelerations. Other insects with heavy abdomens may similarly detect rotation changes. However, there are some quite heavy flies with two wings. They will be able to sense sideways tilts and correct for them by putting more force into their wing-beats on one side. They will also be able to offset any tilts fore-and-aft by turning both wings either ahead or astern, but their problem is to detect these nose-up or nose-down rotations when they get airborne. Otherwise, as we have seen, they will nosedive straight into the ground.

Two-winged flies carry a pair of vestigial wings aft in the form of knobs on the ends of stalks and these vibrate like wings when they fly. If we swing something round on the end of a line and look at it sideways, it appears to oscillate from one side of the hand to the other. In the same way, the up and down oscillations of the vestigial wings have the effect of a rotating gyro-wheel working in one plane. A gyroscope registers rotations and not rotation changes and so these oscillating knobs record fore-and-aft rotations which is all the fly needs. It seems likely that four-winged flies can also use their buzzing rear wings in this fashion but we cannot prove this for, if we remove these wings, they cannot fly.

Self-sensors. We have seen that insects measure gravity, tilt, speed change and sometimes rotation change by the alignments of parts of their bodies, the alignments being sensed by what we may call self-sensors, which tell an animal's brain what is happening to the body when it moves about. We shall use this rather clumsy term, in place of the word proprioceptors normally employed in animal books, because the latter includes all accelerometers and other devices that contribute to stabilization.

When a message is sent down a nerve channel to tell a muscle to move, the brain has to be reassured that what is supposed to happen has happened, and so it uses the feed-back of information from self-sensors. Therefore all animals carry these sensors and insects in particular employ them to provide stabilizing information. But so do other creatures. For example, a fish carries a swim-bladder, a bag generally filled with oxygen, and alters the pressure on this bag by internal muscles, so as to keep itself at a constant depth in the water without having to work all the time with its fins. Self-sensors on the bladder also provide an additional accelerometer.

Vertebrates all use self-sensors and, if they lose the use of a sandwich-sensor, they may compensate by sensing any shifting in the weight of their insides, particularly their stomachs. Mammals carry detectors in joints, on tendons and in muscles, and the messages they send to the brain also enable the animal to remember what it has done. Thus the elephant-shrew, named as a result of its long nose and despite its diminutive size, uses its self-sensors to plant its feet always in the same spots with perfect precision as it escapes to its burrow along a runway. Similarly, human beings use self-sensors to navigate themselves around their bedrooms in pitch darkness.

LABYRINTHS

Any system that measures the vertical by using forces supporting something against gravity will be unable to distinguish tilts from various speed changes without some knowledge of rotation. Indeed, what we call gravity is itself distorted, like the surface of the sea, by speed change due to the rotation of the Earth. We have seen how creatures correct the situation by separate sensors that detect rotation changes. Vertebrates include these with their accelerometers in large complicated sacs known as labyrinths. Fig. 5 shows the labyrinth of a bird, grossly exaggerated in size.

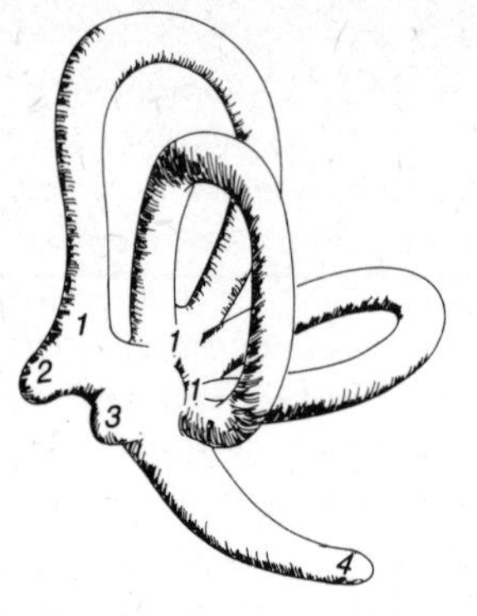

5. Labyrinth of a bird

Unlike the octopus' sac with flaps, a semicircular canal only registers a change of rotation in one plane. Therefore the labyrinth holds three such canals each at right angles to the other two. It also carries a main sandwich-sensor which is mounted horizontally, aligned with the top and bottom of the head. A second sensor, generally below, may be parallel to the sides of the head and therefore upright. A third sensor, usually aft, is also normally upright but aligned across the head at right angles to the other two.

With three sensors at right angles to each other, accelerations and changes of rotation can be felt in any direction, and the vertical detected irrespective of which way up the head may lie. The flatfish provides an interesting example of this flexibility. When young, it swims upright but, by the time it has reached the prime of life, it has turned onto one side and the eye underneath has navigated itself to the top. Not so the stabilizers, which work perfectly well whichever way up the head may be.

Labyrinths are embedded each side of the skull behind the eyes and in front of the ear-holes and, integrated with them, we generally find hearing organs. These also use hairs, but not in jelly, and so they are able to detect the minute vibrations of sound waves rather than the larger discursions arising from accelerations or from swirling liquids. Scientists believe these hearing devices appeared later in the evolutionary process than the balancers, almost as an afterthought. The overall reliance on the bending of hairs becomes less striking when we realise that even short hairs on head, body or hands enable us to sense quite gentle contacts.

In each labyrinth, certain very rudimentary fish only have two semicircular canals and, perhaps suprisingly, mammals have only two sandwich-sensors, the one that is aft being taken over by the hearing organ. A sandwich-sensor can detect accelerations along or across its heap of granules and so the problem may not be acute. In any event, to compensate fully, the mammal can use information from self-sensors in heavy parts of its body, particularly the stomach. Other self-sensors can record the alignment of these organs relative to the head, those in the neck passing information regarding twists and turns.

Differencing accelerometer outputs. A brain capable of sorting out signals from two labyrinths and from self-sensors in neck and body ought to be able to subtract the outputs of one labyrinth from those of the other. In Fig. 6(a), we show a mammal's head from behind so that its left and right sides coincide with the reader's, and with the two lower vertical sandwich-sensors greatly increased in size. In the illustration, the left-hand heap of granules is being pulled further down against the supporting jelly than the right-hand heap.

Obviously, gravity and vertical speed changes will act equally on both sides. It follows that, in our figure, the animal's head must have an anti-clockwise angular acceleration or change of rotation. So the brain can use this information to support the indications of the semicircular canals. Thus an animal will carry a back-up system for sensing 'rotations'.

Differencing has two major disadvantages. First, the two sensors have to be rigidly fixed together, otherwise an apparent change of rotation might be simply due to looseness. It may be noted that, in the human skull in particular, the two labyrinths are joined by extremely hard bone. Secondly, incredibly small differences between accelerations have to be detected. Yet it can be shown by experiments that, if a sandwich-sensor be injured, an animal may rotate itself, thus indicating that these sensors also signal 'rotations'.

Suppose the vertical sensors on the right side of a skull were to be damaged and cease to pass information. The output from the right would become nil but that from the left would continue downwards and, as Fig. 6(a) shows, this would indicate the start of a strong anti-clockwise roll, to be offset by a powerful clockwise correction. Indeed, an animal will fall over towards the injured side.[4] Eventually, the brain learns to compensate, for example by imagining a normal

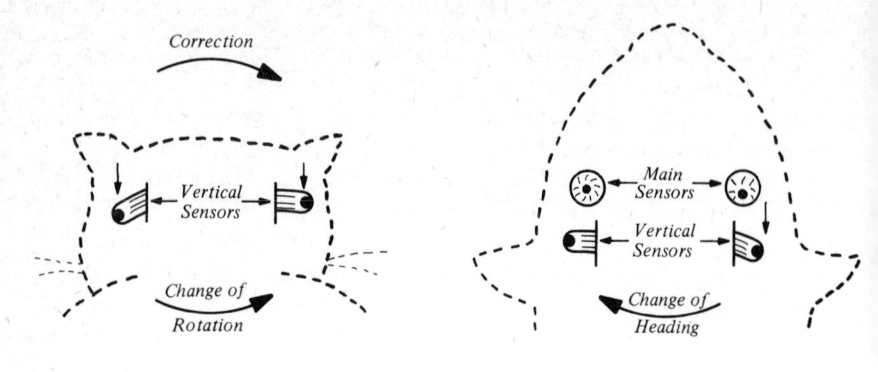

6. Differencing accelerometer outputs (size of sensors grossly exaggerated)

output from the injured side. If now the good side ceases to function, the brain will only receive the imagined signal from the other side, indicating a rolling change towards the sensor originally damaged, to be corrected by a rotation towards the newly injured side. This again will happen in practice.

Thus we can be confident an animal uses the differences of accelerometer signals to support its semicircular canals. Indeed, it seems certain that quite simple creatures without any other means to sense a change of rotation may employ the system. Thus a mollusc, with one sac injured, may roll over towards the damaged side.

Just as differences in acceleration signalled by vertical sensors can indicate the start of a roll, so differences between fore-and-aft sensors will suggest a change in rate of turning. In a mammal, both the main sensors and the lower sensors would combine as shown in Fig. 6(b), in which the rearward tilting of the right-hand sensors indicates the start of a clockwise turning of the head.

Similarly, a difference between the vertical shifting of the lower and the aft sensors could indicate a change in the fore-and-aft nodding of the head, but the distance between the sensors is very short except perhaps in a bird. In mammals, the absence of aft sensors might be compensated by acceleration indications from the weight of body organs, but the links between these and the head may be too tenuous for very accurate differencing. In any event, the fore-and-aft semicircular canal is generally bigger than the other two, so that it can take on most of the responsibility.

LINKS WITH BRAIN

The stabilizers and sensors we have examined measure two types of information:

(a) *Accelerations* from sacs holding granules or sandwich-sensors and from self-sensors. The outputs are an amalgam of
   (i) Gravity,
   (ii) Speed change and
   (iii) Tilt.

(b) *Change of rotation* from sacs without granules or from semicircular canals, supported by differences of accelerometer outputs.

The brain. Fig. 7 indicates how the brain has to combine this information. The outputs required for various purposes are:
(a) Rotation,
(b) Speed change and
(c) Gravity.

The first step must be to summarise or integrate changes of rotation to derive rotations. (This process will not be needed in one plane by flying insects that use buzzing wings which measure fore-and-aft rotations directly.) The rotations will then have to be re-integrated to provide a measure of tilt, which can be subtracted from the outputs of the accelerometers, so that only gravity and speed change are left. This dual signal is then smoothed so as to produce an average direction for gravity, the discursions representing speed changes. The average direction of gravity is, of course, the vertical.

Perhaps a little explanation is required in respect of the smoothing. What the brain does is to base the direction of gravity on the general tilt of the sandwich-sensors in the immediate past, or on the average point in the sac upon which a granule rests, the brain also allowing for any deliberate tilting of the sensors themselves, due to motions of the head. Any discrepancies from this average direction will be regarded as being due to speed changes, tilts having been eliminated by integrating rotations.

The immediate accuracy of the vertical will obviously be affected by the strength of recent speed changes. Thus, when taking off in an aircraft and subjected to strong forward acceleration, the floor may

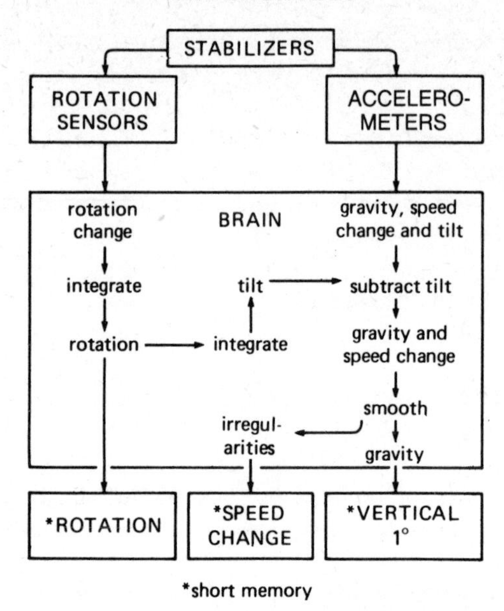

7. Outputs derived from stabilizers

seem to be tilted upwards more than it really is. However, Fig. 7 shows a value of 1° attached to the vertical. What little evidence there is available seems to support this estimate. An error of 1° in a vertical represents a distance of about 70 miles on the ground. Harold Gatty found that Polynesian navigators using the stars could find position from their own verticals to roughly this accuracy.[5]

The wonderful journeys of David Lewis in the catamaran *Rehu Moana* produced similar results. We shall see, when discussing how birds find their ways, that the value seems to be generally supported by tests made on pigeons. Also, bees do not use the Sun for steering when it is less than 2° from overhead.[6] Oddly enough, the size of the stabilizer does not seem to matter much. The agile dolphin does not bother to grow labyrinths any larger than a hamster's. Yet naturally, the value given as 1° can only be an approximation. For that reason, no attempt has been made to indicate on what percentage of occasions 1° is likely to be exceeded.

It is also suggested in Fig. 7 that the outputs have short memories. An animal may need to know what speed change is affecting it at any time, but seems to have no requirement to remember a past value.

Similarly, rotations will only be integrated over a short period in the past for, over a long time, the tilts one way would cancel out those from the opposite direction. Thus an animal is only interested in recent speed changes and rotations.

The vertical is derived from the elimination of speed change and rotation and so it seems fundamental that it likewise cannot involve anything more than a short memory. If it could be retained in the mind for half a minute, it would take this amount of time to become giddy and the same amount of time to recover from giddiness. Indeed, the direction of the Earth's vertical changes by up to 1° every four minutes, due to the rotation, so that animals would lean sideways if their verticals could be remembered for long periods. Thus we may safely assume that the vertical will be remembered for seconds of time rather than minutes. Nor does an animal carry alternative stabilizer systems, otherwise it would not be so helpless if deprived of the sensors we have described.

Links with senses. Nevertheless, an animal uses other senses to support its measurement of the vertical. One has only to stand on one leg and shut the eyes to realise the part vision plays in keeping upright. Many very rudimentary creatures have light sensors on their backs or in the tops of their heads. The caterpillar of the clouded-yellow butterfly has light sensors on its back and, lying along the upper side of a leaf-stalk, will move round and cling upside down if lit by a lamp from below.

A brine-shrimp in an aquarium which is illuminated from one side similarly leans over at right angles. A fish-louse with a light underneath turns a somersault. If a prawn has lost its sacs, it may still rely on the pressure of the ground against its feet and on illumination from above, though it will be distinctly unsteady. Light it from one side and, as the two sources of information conflict, the crustacean settles the matter by leaning through 45°.

Discrepancies between eyes and stabilizers have dramatic effects on human beings. If we sit in a railway carriage and a train next door starts up, we feel we are moving the opposite way and, when the train has just gone, we seem to decelerate suddenly and stop. In a ship's cabin when the sea is rough, the eyes may assert the surroundings are not moving, but the sandwich-sensors and semicircular canals tell a different story, and so does the stomach supported by other organs inside the body. The result is confusion of the system, a sense of untoward motion within the stomach and consequent malaise and sickness.

Nevertheless, in general, the eyes render a great deal of assistance by being able to detect very small changes of alignment. Sensations from the eyes pass to the conscious or thinking part of the brain, which may transmit a message to a hand to move to somebody or something for support. At the same time, information goes to the unconscious or instinctive brain and contributes to the 'feeling' that the body is or is not upright.

The power of the eye to reveal the truth or to deceive is remarkable. Professor H. A. Witkin of New York built a chair into which a willing student could be seated, the front view being filled by a representation of a room complete with door and pictures. This 'room' could be tilted sideways through an angle and the chair could likewise be tilted through an angle, but under the control of the victim. In the experiment, the lights were turned out, room and chair tilted and the lights then turned on. The victim was then asked to sit in the attitude he would adopt if eating a meal. It was common for both chair and room to be tilted through large angles and for the victim still to believe he was upright. 35° was not unusual and the record was 52°! The victim was then asked to close his eyes and at once the tilt would be sensed by his now uninfluenced labyrinths.

Not only do eyes confuse stabilizers but also stabilizers may complicate matters for the eyes. We have already described how a crab's eyes react to its 'rotation' sensors. The same effect happens with higher animals and with human beings. When a spinning is established, the liquids in the semicircular canals rotate at the same rate, the sensitive flaps become centralised and the eyes cease to react. If the spin is then stopped, the eyes will fix on a stationary object but the flaps, pushed the other way by the swirling liquids in the canals, register a violent angular acceleration opposite to the original reaction. The consequence is giddy confusion.

By means of stabilizers, animals keep their heads steady when they are moving about. Films taken of birds show they keep their heads level even though gusts of wind make their wings dip and turn. A bird sitting on a branch swaying in the breeze likewise maintains its head steadily aligned. Even when walking about, though apparently nodding, it actually keeps its head steady, pushing it forwards with

3. *Green tree frog*
Certain tree frogs use their verticals to lay eggs in trees so that the tadpoles fall into water when they hatch.
Gary Weber, Aquila Photographics

Bird landing on a branch. This rook uses its accelerometers to control its landings. *M. C. Wilkes, Aquila Photographics*

1. *Crane fly*. The two-winged fly carries a pair of vestigial wings
form of knobs on the ends of stalks. These halteres act as
gyroscopes. *Ron and Christine Foord*

2. *Termite queen* being tended by workers. (Soldiers, with larger he
also seen.) Termite queens lie across the Earth's magnetic field.
Bannister, Natural History Photographic Agency

the neck and then letting the body catch up. One type of frog uses its vertical to lay eggs in a tree twenty feet above a puddle so that, when the eggs hatch out, the tadpoles fall neatly into their preferred environment!

Fig. 8 emphasises the connections between brains and senses. It shows sight passing information to the conscious brain, though in fact it is routed through an unconscious sorting office, from which the brain decides what it wants to look at. The unconscious brain also takes signals from the stabilizers and amalgamates all these into instructions to the eyeball muscles to move the eyes. The results pass back to the sorting office in the brain but, for simplification, are shown as dotted lines leading to sight and to the stabilizers.

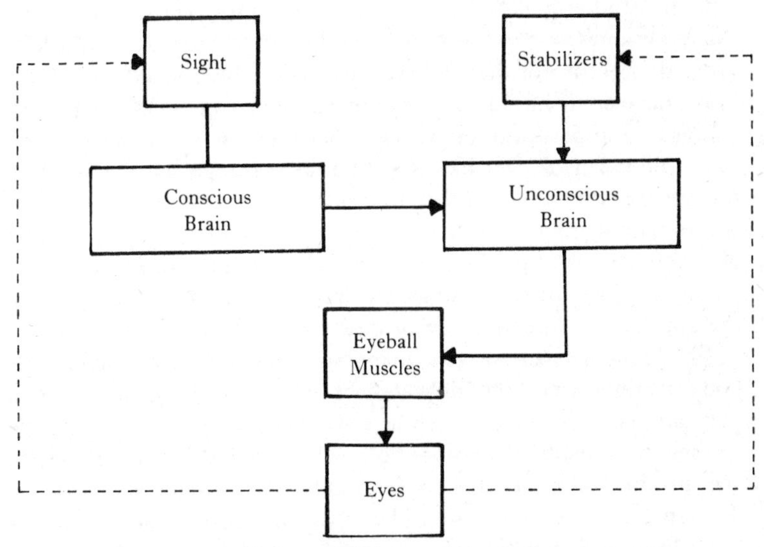

8. Sight, stabilizers and the brain

We can prove that eyes are controlled in this way by a very simple experiment. Shut the eyes and put the fingers of both hands gently on the lids, holding the palms against the chin to keep the hands steady. Then, deliberately using the conscious brain, turn the eyes from side to side and the movement will be felt by the finger tips. Next, repeat the exercise without any deliberate turning of the eyes but, this time, twist the body to left or right and the eyes will be felt to turn instinctively to right or left so as to keep 'looking' in the same direction. They do this even when we are asleep. However, a dancer learns to defeat the system during a pirouette by flicking the head round quickly and then holding it still for a moment, thus offsetting swirlings in the canals at the start of the rotation by opposite swirlings at the end. This is a process similar to that used by very simple creatures that dart about in order to balance out their speed changes.

Although we may turn our heads or move our eyes so that we look at objects in various directions, our unconscious brain will reassure us that things fixed in our surroundings 'look' as if they are not moving even though our eyes are travelling across them. However, if we deflect the eyeball without using the brain to control it, the result is different. Look at something stationary, shut one eye and move the other by prodding the lower eyelid gently upwards with a finger. The thing we know is fixed appears to leap downwards. It would leap upwards were it not that the eye turns everything upside down.

We have concentrated on vision because it is the sense we human beings best understand. However, other sources of information need to be stabilized. A deer listening to a noise rotates its ears and tilts them to find direction and then moves its eyes to look at what it hears. Animals may rely on other senses, such as infra-red, scent and electricity, to paint pictures of the world around them. All these rely on the stabilizers acting through the brain for translation into action in the appropriate direction, up or down and left or right.

The brain of an animal also links stabilizers to muscles that keep its body upright. Yet the delicate act of balancing is no more a navigational requirement than legs are navigational equipment. We can see this must be so because both balancers and legs are needed to stand quite still or to jump up and down at the same spot. Just as self-sensors have uses far wider than navigation, so stabilizers contribute to locomotion as well as to its purposeful control.

Watch a tiny tropical fish nosing about its tank, see a wren flutter across the garden, glimpse the sudden movements in the cornstalks that betrays a field mouse. Each of these minute creatures has, in the sides of its skull, a pair of labyrinths with semicircular canals and sandwich-sensors and, in the millimetric space between, a brain analysing the signals instantly and relating them precisely to sensations from other parts of its body.

TIME

The ability of animals to sense time is not only widespread but also appears to be embedded deep into the past. One crustacean, which probably appeared on this planet a hundred thousand years ago, now lives in caves at the bottom of deep oceans where there is no light and where the temperature never varies. Nevertheless, it still retains traces of regular changes each twenty-four hours[7] though the Sun would only distort gravity by about 1/1,000′.

Where the universal sense of time comes from remains a mystery. Sometimes it must be due to some external influence, and the creature may rely on senses about which we know nothing. At other times, it seems to be connected with the basic processes of living. The arrival times at which bees come for food may be put forward or delayed by drugs which speed up or slow down the metabolic rate at which the insects live. Other rhythms may be linked in vertebrates to rate of heart beat or respiration. Whatever the method may be, the capability appears in single-celled creatures as well as complex mammals and seems virtually universal.

Many rhythms are of quite short duration. That creature greatly prized by sea anglers, the lugworm, operates to a forty minute schedule, coming up tail first to produce a worm cast and then pushing out its nose every seven minutes to take in sand and, hopefully, bits of edible matter. It will continue to go through the motions even without sand and with large portions of its anatomy absent. Moles, shrews and mice split their days into exact periods of activity and of sleep, generally three or four hours. Some birds repeat songs with a timing accuracy of a fraction of a hundredth of a second so that two such birds may appear to be singing in perfect unison. However, as a general rule, timing rhythms that affect navigation are, as shown in Fig. 9, either:

(a) Daily,
(b) Tidal or
(c) Yearly

or sometimes a mixture of these.

Daily Rhythms. In many instances, the internal clocks that keep daily rhythms are linked to phenomena such as dawn or dusk which serve to put the clock right if it starts to slip. Certain bats hunt for two

RHYTHMS	SUPPORTING INFORMATION
Daily (Order of error 0.1 hours)	Sunrise and Sunset
Tidal	Motion of Water
Yearly	Length of Day

9. Timing rhythms

hours at dusk and two hours at dawn and, as their temperature falls when they sleep, their alarm clocks have to rouse them some time before these conditions of light appear, so that they may warm up before flying.

Cockroaches likewise come out at dusk, but their rhythms can be altered by artificial lighting. Also, if kept in continual darkness, they move about for the whole twenty-four hours, though naturally rather sluggishly. If the body liquid from normal dusk risers is injected into these cockroaches without rhythms, or even if a light is flashed at them at sunrise, they take up their normal daily rhythm again. The timing is controlled by four cells in the brain and, if these are removed and replaced by those of another cockroach with a different period, this new period will then control the insect.

A fruit-fly, the size of a v in the small print of a hire purchase agreement, hatches out in the cool, damp air of dawn so that its body can harden and its wings untangle before the heat of the day dries them out. When produced from eggs and larvae in complete darkness, a clutch will have nothing by which to set its clock and so the flies emerge at random. Fire a single thousandth of a second flash at them a few days before they are due to arrive, and they take this for dawn and all pop out more or less together some multiple of twenty-four hours later.[8]

Certain twenty-four-hour rhythms affect plants as well as animals. For example, flowers have opening times by night or by day according to which customers they are designed to attract for pollination. Transport a hive a long distance east or west and the bees continue to visit the flowers at the original times, not realising the local clocks are ahead or behind those which they have been using. However, the bees soon learn to adjust their visiting hours.

The ways in which animals react to drugs, poisons and various stimuli also vary with the time of day. At the same time in each twenty-four hours, certain sounds will convulse mice fatally which, at other times, they completely ignore.[9] But perhaps the final accolade should be given to the humble potato. Its internal processes are said to have a rhythm four minutes less than 24 hours, the exact time it takes the Earth to complete one revolution in space!

In any event, the regular habits of animals are well known to everybody. Fish can learn the times at which humans feed them. Birds visit gardens at regular hours. Dogs go to railway stations to meet their masters returning from work on trains, their timing being far more reliable than that of the trains themselves. Ordinary human beings can teach themselves to wake up at certain times and may even be able to change these times at will. In spite of senses dulled by clocks and watches, they can often tell the time to within ten minutes, particularly if they are of regular habits.

From sources such as these, it may be possible to suggest the order of accuracy of animal timekeeping. Reports of experiments with the hatching of fruit-flies, the timing of cockroaches and the behaviour of other lower creatures suggests that a fraction of an hour each day may be achieved, possibly reducing to ten minutes in certain instances. The capabilities of birds and mammals seem to be even higher with errors perhaps coming down to a few minutes a day, say 0.1 hours which is probably the 'order' of the error. 0.3 per cent has been given by C. Rawson which works out to 0.072 hours a day.

Timing errors per day may become less over a period, the systematic or constant slippages being multiplied by the number of days, but the random variable elements being reduced by a factor according to the square root of the number of days. Thus, if an error of 0.1 hours a day was built up from 0.4 hours of systematic slippage and 0.06 hours of random elements, after four days the total error would be $0.04 \times 4 + 0.06 \div \sqrt{4} = 0.16 + 0.03 = 0.19$ hours, an average of about 0.045 hours per day which is less than half what it was in the one day.

Tidal Rhythms. The rhythms of creatures that live near the shore naturally coincide with the tides and are repeated every twelve hours and twenty-five minutes. The little fiddler crab changes colour at high tide to camouflage itself, and will continue to act as a reasonably reliable tide predictor for long afterwards if placed in an aquarium.

Even more remarkably, if we nip off a piece of skin and feed it with nutrients, the skin will continue for a time to change colour with the daily tides.[10]

A limpet may go for a walk when the tide comes in but will be back at its place on the rock before the tide goes down. Oysters open and shut according to tidal rhythms even after they have been taken out of tidal waters. It has been suggested that they change these rhythms to fit the local times of high and low water when taken a long distance east or west from their home district,[11] which implies an ability to sense the Moon. The oysters in one experiment were not subjected to moonlight and the gravitational effect of the Moon would only shift the vertical by about 1/500th of a minute. We can only assume an influence exercised by the Moon about which we have as yet no information.

Yearly Rhythms. Many of these may be connected with simple annual phenomena such as the shortening of the day in the autumn and its lengthening in the spring. A weasel can be persuaded to keep its summer coat by increasing its daylight hours artificially using electricity. But some of these rhythms are certainly inbred. Dr. E. F. Sauer has found that those attractive little migrating birds, warblers, hatched and hand reared in soundproofed rooms, artificially lit all day and night and maintained at a constant temperature, become restless at the time they ought to migrate but settle down after the period it would take them to fly to Africa. In the spring, the restlessness is repeated throughout the time the warblers would be flying back from Africa.

Some rhythms are a combination of tidal and yearly. The grunnion is a little six-inch fish that, on spring tides between March and June, lays its eggs as far up a sandy Californian beach as it can. The eggs are apparently affected by being shaken by the next spring tide a fortnight later, for the young emerge from their eggs to welcome it and then swim out to sea to grow up and repeat this remarkable performance.

The palolo worm sheds its tail at dawn, exactly one week after the November full-moon, and then proceeds unconcerned on its way. The tails, each with one eye, rise to the surface to mate with other tails and are harvested by the local Polynesians who enjoy the annual feast provided so regularly.[12] These are but examples of the many extraordinary ways in which animals manifest their timing abilities.

SUMMARY

1. *Stabilizers*. When processed by the brain, these record

(a) *The Vertical*. Supported by sensors, particularly eyes. Accuracy probably around 1° in higher animals.

(b) *Rotation*. Supports sensors such as eyes.

(c) *Speed change*.

The outputs have memories probably measured in seconds rather than in minutes.

2. *Time*. A universal sense, registered by higher animals to the order of about 0.1 hours in each 24. Constant slippages increase according to the number of days, but random variations decrease according to the square root of the number of days.

2

Motion

LINKS WITH STABILIZERS

Engineers may find it convenient to divide motion from place to place into two velocities at right angles to each other. However, animals have a direction in which they can best advance or retreat and, as they are normally symmetrical, this will generally be the line from head to tail. So we shall relate motion to speed and to the direction of this speed, as shown in Fig. 10.

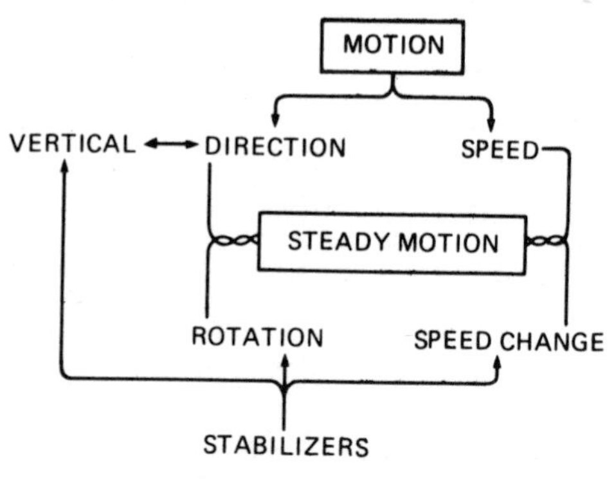

10. Motion and stabilizers

Of the two components, direction is much more important. If a creature does not travel the right way, it will not get there. All speed can do is to make it arrive sooner or later, and animals in their natural state do not have to keep to schedules. Nevertheless, it may be necessary to travel at a steady speed, though an animal may have to call on maximum effort if it needs to catch prey or has to avoid being caught.

Speed change. To achieve a steady speed, it is essential to know when this speed changes. For example, suppose a fish is swimming in a shoal or a bird flying in a flock. If it finds it is falling behind the others, it has to make up the gap by going faster. When it has caught up, it will be travelling faster than its companions and so it will forge ahead. For, in the absence of a knowledge of speed difference, it will be impossible to know that speed is wrong until a discrepancy in distance has built up. The animal then has to reduce this overshoot by going slower and eventually it lags behind. Thus its position will oscillate between ahead and astern compared to its fellows.

If the creature can measure speed change, which is sensed by the stabilizers, the problem disappears. As it begins to catch up after the initial lagging behind, it can slow down according to the distance it is behind until, when it is level, it is travelling with no speed change compared to its companions. Furthermore, if the shoal or flock should start to forge ahead, by detecting speed change the animal can accelerate to keep level before an appreciable gap has built up between itself and the others. Thus a measure of speed change, generally supported by evidence of the eyes, is an element needed for the steady control of speed.

Speed change is also needed if an animal wants to stop at a certain place. Consider a bird wishing to perch on a branch. It needs to be travelling at close to zero speed when it lands but cannot do this by simple alterations of speed. For if, when nearly there, it is travelling at zero speed, it will not reach the branch and, if it is not, it will overshoot. So the bird has to learn to change speed, steadily reducing it as the distance to the branch closes and, from the way this speed change is related to the distance still to go, it arranges that the forward motion will have died away completely when it arrives. During the whole of the approach, the bird is able to correct the way its speed changes. So it can land on a branch unaffected by the wind and irrespective of whether the branch is swaying about at the time.

Rotation. When a young baby goes to touch something just out of reach, it moves its hand towards it by rotating its arm and, on arrival at the right place, the hand is still moving. It overshoots and the baby sees this, slows the hand down and once more rotates the arm towards its objective. Once again, the hand moves across the target and overshoots on the other side. Thus the baby waves its hand to and fro across the object it wishes to touch.

After a time, the baby learns the knack. It slows down the rotation of the arm as the distance of the hand from the object reduces until, on arrival at the target, the hand has come to rest, subsiding at the correct position and no longer oscillating to and fro. We notice the baby takes great satisfaction in bringing its hand to a stop at exactly the right point to pick something up. This demonstrates the need for a measurement of rotation in order to correct a direction and explains why the use of rotation to support direction, and also of speed change to support speed, are processes known to engineers as 'quickening'. In addition, the change of rotation sensed by stabilizers can 'quicken' rotation information.

The vertical. In Fig. 10, links were shown between direction and the vertical. For direction, we may consider a very simple example and suppose we wish to walk to a tree on the other side of a valley. Immediately we notice there is a pond on the direct line. Although this is below the level of the tree, we can tell it is on our route. One thing is certain, the pond must lie on a vertical through the tree as seen by our eyes. This tells us how we identify our path towards anything. We simply imagine a vertical through it.

The next problem is to steer straight towards the tree or the edge of the pond. Human eyes, like animals' eyes, are very sensitive to anything moving at the sides of the field of view for this is how the presence of prey or a predator is noticed, even when looking at something else. So, as we walk towards the tree, we can see at once without looking whether the ground close to the feet is drifting to left or right of our imagined vertical. If it drifts to the left, we must be going too much to the right so we veer left. Thus we instinctively alter course until, as we pass over the ground, we are aware it is travelling straight down the vertical through the tree or the edge of the pond or wherever we want to go to.

This is, of course, how an animal 'homes' when it can see something, and it is illustrated in Fig. 11(a). It is possible to work without a vertical by carrying on irrespective of whether the path is right or wrong until it becomes obvious that the direction of the objective is shifting. This may take a long time unless the object is close by, but it may be the only way to manage if sight is not available and some other sense has to be employed. For eyes have the wonderful facility of detecting, in addition to direction, movements of the ground below in the form of changes of direction, which can 'quicken' the purely directional information.

It should be recognised that the object to be reached need not be at the same level as the animal, for the vertical through anything can be sensed even if it be above or below the eye. Thus an animal may walk towards the Sun in the sky or towards a lair in the valley below. By its vertical, it may even be directed by a sloping line of magnetic force, irrespective of whether this dips downwards or is inclined upwards.

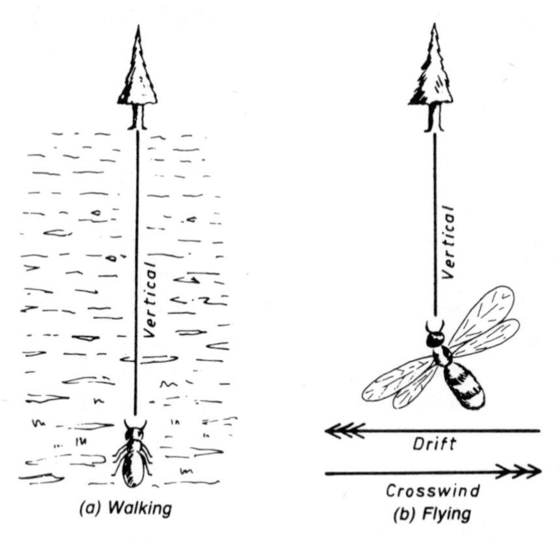

11. Using the vertical

The vertical can be used in exactly the same way by a creature flying in a cross-wind. If the wind is from the left, as in Fig. 11(b), it will blow the bee or bird to the right, and the ground below will *appear* to drift to the left. Once again, the animal will offset this by altering course to the left until the ground ahead and below is travelling down the vertical through the objective. It will be no more difficult than following a vertical towards an object when walking on the ground.

An aircraft pilot may well quarrel with this last statement. However, when landing visually, he has to manage to crab by the correct angle so as to offset a cross-wind, even though the nose of his aircraft is obscuring the ground ahead and below which would give him directly the information needed. He may have to look out of a side window and visualise, from the shifting patterns he sees, how the ground underneath is drifting, a remarkable feat. The pilot of a glider with less obstruction ahead finds it easier. The old-fashioned bomb-aimer, using a crude sight but looking down at the ground directly from above, was hardly aware of the problem. However, even he had difficulties over the sea.

In light winds, the absence of 'white horses', as breaking waves are called, makes the measurement of drift impossible. In strong winds, the overall pattern of the sea is confused and continually changing and the motion of individual white horses has to be established. That this is beyond the ability even of seabirds is evident from the fact that their paths are deflected by prevailing winds.

Interception. Anyone watching a wild-life television programme which shows a cheetah chasing a gazelle will be struck by the way the hunter runs in the exact direction at which to intercept the fast-moving prey. A cheetah, moving straight towards a gazelle fleeing at right angles, would have to change course progressively and travel along a curved path which would degenerate into a chase from behind. Any increased distance to be covered by the hunter could only benefit the hunted, and improve its chances of escape.

All the hunter does is to travel as straight as it can towards the moving prey allowing for the fact that the prey is itself moving. It does this by using change of rotation sensors to keep the direction of the quarry as constant as possible. In Fig. 12, we see what happens when the prey alters its path but the same thing would apply if it were to accelerate or even to slow down. The widely spaced arrows show how the direction of the hunter's line of sight to the quarry will begin to shift.

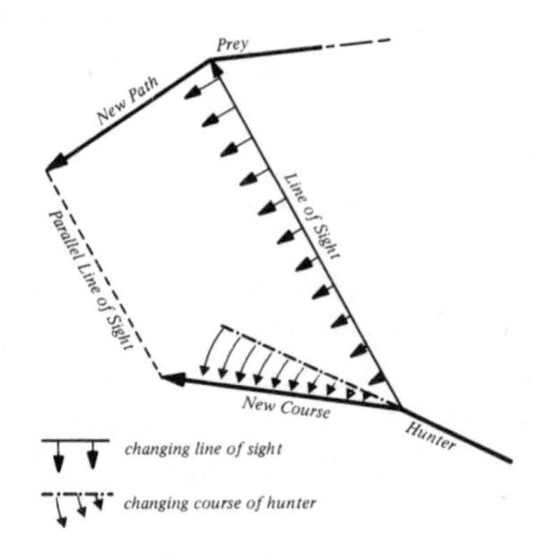

12. Proportional navigation

If the hunter were merely to alter course to follow this drift in the line of sight, the drift would continue. So the hunter alters course faster, as shown by the closely spaced arrows, and so 'kills' the drift in the line of sight. As soon as the drift is killed and the line of sight steadied, the hunter steadies its course and the consequence is a perfect interception. This method of disproportionate correction has been adopted by engineers who design anti-aircraft missiles and is known to them as 'proportional navigation'.

The nearer a hunter approaches to its prey, the greater will be any change in the line of sight resulting from an alteration in course or speed of the quarry. A point will be reached just before the hunter strikes when a jink by the prey will produce a shifting of the line of sight faster than that which the hunter can offset. Such a jink has to be left until the last possible moment so as to produce the maximum possible rate of directional change. This instinct is common to all animals including human beings. Mariners at sea tend to alter course too late when avoiding a collision and have to be specifically trained to turn early and, in order that the other craft can see what has happened, by a large amount.

In all that has been said about direction by visual means, it is worth noting that most animals can see everything that comes into their eyes simultaneously, without having to turn them to look at each particular thing, as humans have to. Furthermore, they can generally see large areas of ground, particularly ahead, and so have no difficulty in following a direct route defined by a vertical. Nor do humans have any problem, for our eyes detect motion in all directions even when we are looking directly at other things. Also, most animals have eyes that can see the whole sky and so they can simultaneously watch the Sun or the Moon and stars while looking at other things.

USING SUN, MOON OR STARS

Sun. Very primitive land creatures may move towards the Sun by sensing the heat or light on the fronts and tops of their heads and bodies. The water-flea, when frightened, scurries over the surface of its pond towards the Sun, the level water providing vertical information and the manoeuvre probably tending to carry it to the shadow of a bank, a feature that presumably enabled a greater proportion of its ancestors to survive and breed.

That fascinating creature, the dung-beetle, is able to alter course by 45° every three hours[13] and so travels roughly in one direction but along a series of arcs. To alter course in this way, the dung-beetle has to change the position of the Sun from being dead ahead to being halfway between ahead and abeam, and later it has to keep the Sun on one side, making these changes according to the time given by its internal clock.

L. Pardi and F. Papi have studied the little sand-hopper which likes to live on beaches in damp sand and is too low on the ground to see far. Suppose the beach faces eastwards. If the hopper feels the sand is too dry, it moves eastwards towards the sea and, if it is not dry enough, it travels westwards away from the water. Should the same hopper be moved to a westward facing beach and the sand is too dry, it hurries east, not knowing this is landwards, but hastens westwards out to sea if it feels too wet.

By putting sand-hoppers in dishes, the experimenters found the little creatures move in the correct direction only when they can see the Sun and, if it is reflected by a mirror, they move in the opposite direction. Furthermore, like the dung-beetle, they are using their clocks. It was found that these timepieces would stop if the hopper was too cold and, when stopped by six hours, its paths were generally 90° in error. So it must have been using time!

Many other simple creatures, spiders, locusts and ants for example, orient themselves by the Sun using their sense of time and we shall discuss the remarkable abilities of bees later. By reflecting the Sun, it has been demonstrated that salmon and other fish have similar capabilities. Sunfish, all of whom live in the northern hemisphere, can cope with an artificial sun in the form of an electric light, provided it moves from left to right as the real Sun would. Cichlid fish, which live on both sides of the equator, are limited rather differently. An artificial sun has to move in the same direction as the real Sun did when the fish was young, otherwise it cannot employ it.

Amphibians also orient themselves by the Sun and, among the reptiles, so do lizards and turtles. Dr. Gustav Kramer was the first to demonstrate the ability of birds. Amongst other experiments, he changed the internal clocks of starlings by exposing them for four days to artificial daylight differing from the real daylight by six hours. The birds subsequently produced errors in direction of around 90°, as would have been expected with timings altered by six hours. Yet relatively little work has been undertaken with mammals, except for small rodents such as wild mice which also seem able to orient themselves with reference to the Sun.

One particular problem is apparent when finding the direction of the Sun with the aid of time. A sundial has its hour marks unevenly spread over its face because it has to convert the steady motion of the Sun across the sky into changing directions in the horizontal plane. A navigator has the same problem when plotting the azimuth of the Sun on his chart and has to resort to tables or to a computer.

The reason for this appears in Fig. 13(a), a diagram intended to represent a human observer looked at from above with the horizon depicted as a surrounding circle. Assuming we are in the northern hemisphere, the Sun rises in the east and sets in the west but, as it is millions of miles away, it has to be represented by a dummy Sun close to the observer but changing its direction as the real Sun does. Thus the path would tilt upwards, at an angle of 60° to the paper in this illustration, and every two hours the Sun would move across the sky through an angle of about 30°, giving two-hourly positions shown in the figure.

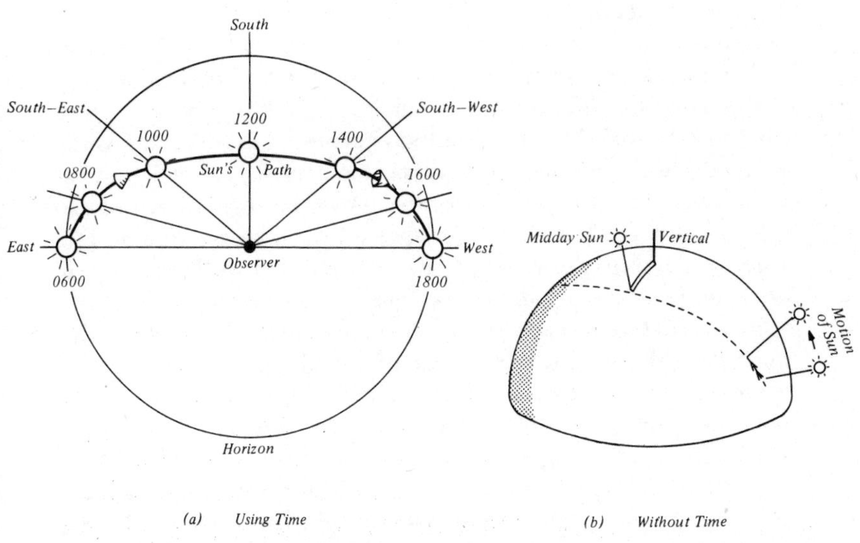

13. Direction from the Sun

To the observer, the direction of the Sun in the horizontal plane appears to change as indicated by the radial lines. At 0600 hours, it is east, at 1000 hours south-east, at 1200 hours south, at 1400 hours south-west and at 1800 hours west, which happens in the spring and in the autumn at 30° north latitude. Such varying changes of direction are often depicted in animal books but may be sidestepped by animals themselves. They only have to follow the steady motion of the Sun across the sky and, if interested in horizontal direction, for example to steer a course, they can use the vertical through the Sun, just as we have done in the picture.

Nevertheless, Professor K. Schmidt-Koenig has stated that birds obtain compass directions from the Sun by allowing for its diurnal motion in this awkward horizontal plane.[14] In the same book, he mentions that some animals find it more convenient to orient themselves by shadows. These are extraordinarily compelling seen from the air and also only change direction slowly at dawn or dusk. This may be an interesting example of how animals may learn something, in this instance the use of horizontal shadows, which they find useful to them in their everyday life even though it may seem complicated to mere human beings.

Is there a need for time? A diametrically opposite solution to orientation by the Sun has been pôsed by Dr. G. V. T. Matthews, recognised all over the world as an authority on bird navigation. Fig. 13(b) shows the dome of the sky and the path of the Sun across it, marked by a solid line. Dr. Matthews believes a bird has the visual acuity to see the Sun actually moving across the sky and indeed it has been said that crabs can do likewise. Also, he suggests a bird can extrapolate the Sun's motion, as shown by the pecked line, to find where the Sun passes closest to its vertical.[15] This will be the local 'midday Sun' and its direction, taken as always from the bird's vertical, will be either due south or, in the southern hemisphere, due north, and this is shown by the double solid line.

One might ask why a bird should be so interested in the highest point something will reach in the sky. All creatures are vulnerable to attack from above for they cannot use their limbs to defend themselves. A fledgling on its nest is greatly disturbed at the sight of a hawk, and will watch closely lest it approaches a point directly overhead from where it can stoop down for the kill. Indeed, for danger from above, many birds utter special warning calls, quite different from the ones they employ for a predator on the ground.

Of course the movement of the Sun across the sky is almost entirely due to the Earth rotating in the opposite direction. Thus the Sun seems to travel across the sky in a straight line in spring and autumn but slows down by rather less than 10 per cent in midsummer and its path curves slightly. It repeats the process in midwinter, curving the other way. However, these deviations would not have a serious effect whether the bird was finding direction by time or by using the midday Sun.

Moon. Although travelling at the same sort of speed as the Sun, the Moon's timing shifts by six hours each week giving errors around 90°, and the speed along that path and the curvature changes in a month as much as the Sun's does in a year. Furthermore, the Moon may not be up when the animal starts out or it may sink below the horizon before the night is over. It seems impossible any animal could find direction from it with the aid of time. Yet some simple creatures seem to use it.

On a damp moonlit night, the little sand-hopper may creep out of its hole and travel a hundred yards or so inland and then return. Ants can also steer by the Moon. Dr. J. Reimann of Freiburg has trained them to orient themselves by an electric light bulb. Provided it travels steadily, the little insects can learn to use it and they manage perfectly well even when it travels the wrong way![16] Hoppers too manage to use the Sun when carried from the northern hemisphere into the southern even though, as seen by the hopper, the Sun will move from right to left instead of from left to right.

Human beings can do the same trick. If we note where the Moon rises and then wait for a while, we shall see the direction in which it is moving. Extend this line to the point where the Moon would be highest in the sky and that point will be north or south according to our latitude. The idea that a line not parallel to the horizon may also be straight should not surprise us. If we could imagine ourselves to be at the centre of a globe, the curved lines of longitude would all look straight to us.

An animal may even use the direction in which a new Moon 'looks', or any other imaginary line dividing the Moon into two equal halves, assuming the Moon is not full. Such a line will be a quick guide to the direction of motion of the Moon and could suggest the highest point in the sky the Moon would reach that night, which of course will be on the true north–south line. It may be noted that, when the Moon is full, many night travellers, including some birds, will not set out, perhaps for fear of predators.

Stars. Animals also use the stars, which travel steadily across the night sky, their various paths painting a picture of the whole of the sky slowly rotating. The line along which stars move neither up nor down is, of course, the north–south line. The Pole Star is along this line and appears not to move at all and the eyes of birds can watch it perfectly well even when flying southwards. Yet blacking out the Pole Star in a planetarium does not destroy orientation though it may make it more difficult. Nocturnal migrants can manage provided they can see the central part of the sky above.

One would naturally expect birds to prefer some convenient star to mark the line of highest points in the night sky. This could be the Pole Star in the northern hemisphere or some point determined by the Southern Cross when south of the equator. Indigo buntings, in a planetarium with the night sky made to rotate about a star in the tropical constellation of Orion, learnt to use that star instead, retaining the habit for a time even when the sky was returned to its normal rotation. It has also been suggested that night moths may use the stars, for their eyes happen to be well adapted to starlight.

As it is possible to find direction from objects in the sky without using time, one might ask why so many animals use their internal clocks to obtain direction from the Sun? Perhaps estimating the highest point in the sky is rather a long and laborious process. By learning how the direction changes during the day, a straightforward and quick answer will be available. In any event, it may not need to be checked against the highest point, for we shall find that most creatures can sense the Earth's magnetic field. As with 'rotation' sensors, this may be yet another example of how animals like to carry back-up systems.

The emphasis on orientation may seem strange to a human being to whom it is only a form of Saturday afternoon amusement. To a creature such as the little sand-hopper, it is a matter of life and death. It is also of overriding importance to a colony of bees who have to make vast collections of minute quantities of nectar from individual flowers to feed themselves and their growing brood. They have to concentrate effort in areas where there is plenty of nectar, and cannot afford to forage far and wide across fields where there is only grass, or trees not in bloom.

Bees. The first requirement is to be able to return to the hive. To get back to his boat, a skin diver may travel along a line determined by his compass, deviating only by a small distance to either side and then going back to his original line. Similarly, the bee flies on a bee-line using the Sun, which it observes through its compound eye. This eye is composed of thousands of identical single eyes packed together, each eye-cell looking in a slightly different direction, though with considerable overlap.

On its initial exploratory flights, the young bee learns the way the Sun alters its direction in terms of one single eye-cell in a certain time interval. The bee can then fly at any horizontal direction to the Sun's vertical and can make subsequent adjustments to allow for the changing position of the Sun. By thus defining the direction visually, a bee can also allow for any drift due to cross-wind. The question is, how does it find the initial direction of the nectar area compared to the Sun?

For the answer, we turn to the magnificent work of Karl von Frisch.[17] He showed that a bee, returning from a nectar area, will dance a figure-of-eight pattern either on the alighting board of the hive or, more commonly, on an upright honeycomb, waggling all the time from side to side. The direction of the central line of the 8 compared to upright shows the direction of the nectar area compared to the Sun, this direction being continually amended to allow for the way the Sun travels across the sky. The other bees jostle round the dancer and touch it with their antennae. As each fresh bee learns the angle, it sets out, aligning itself according to the Sun and flies off making the usual allowance for the Sun's motion as it travels to and from the nectar area. Because correcting for cross-winds is so natural a process, bees make no allowance for this in their waggle dances.

When the pilot of a light aircraft overflies an airfield on which he wishes to land, he may return to the landing strip by flying a 'race track' pattern as shown in Fig. 14(a). A bee may follow similar procedures when searching to either side of the bee-line. This would explain why, on finding a profitable nectar area, a bee performs the figure-of-eight dance as shown in Fig. 14(b).

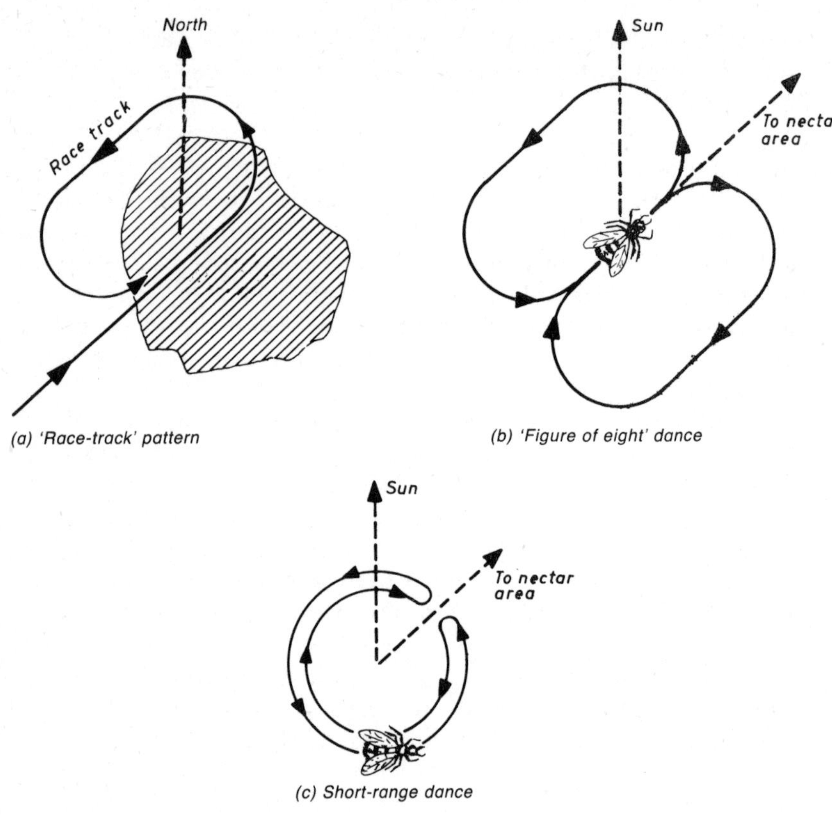

(a) 'Race-track' pattern

(b) 'Figure of eight' dance

(c) Short-range dance

14. Waggle dances of honey bees

If the nectar area is so close it can be found by smell without a search, the bee dances a pattern shown in Fig. 14(c), which does not include any suggestion of a search pattern, but again allows for the motion of the Sun. If the nectar area is even closer and can be found without any need to know direction, the bee may circle rather aimlessly on the honeycomb. Bees do these dances not only to locate profitable nectar areas but also to identify suitable nesting sites when they swarm.

It is remarkable that a bee is able to cope with a bend in a course and yet estimate a mean direction. For example, if the nectar area is on the far side of a high hill, a bee will diverge from its bee-line, travel round the hill and pick up the original path on the far side. Moles do the same if one of their runs is blocked. Also, if an obstruction is placed ahead of a solitary wasp dragging prey to its nest, the insect will turn half-left or half-right, go past the obstruction and then turn back onto its original line, and it will repeat the process if there are a number of obstructions. However, this is a far cry from an ability to fly hither and thither at random, and then convert all the multiplicity of courses, speeds and times into an integrated total direction.

Polarised light. Bees are busy creatures and do not like stopping work every time the Sun goes behind a cloud. Fortunately, their eyes, being small, can detect the very short ultra-violet light waves which penetrate cloud and this allows them to see the Sun when mere human beings cannot. Sometimes the cloud is too thick even for ultra-violet. If there is a small patch of blue sky, the bee can detect what is known as 'polarised' light from the Sun.

Normally, light waves oscillate in all directions at right angles to the path along which they travel. However, when they pass through the atmosphere, many of the oscillations are thinned out into one plane and these lateral waves enter each single eye-cell of the bee. They go through the lens and pass down eight incredibly thin tubes surrounded by eight cells, each of which carries cross-sensors so fine they can be seen only under an electron microscope.

These infinitesimal sensors measure the direction of the lateral oscillations and, mixed with similar sensors in adjacent eyes, they tell the bee's brain the direction of polarisation and hence the direction of the Sun. However, ambiguity can arise when there is only a single patch of blue sky, unless it happens to be roughly opposite the Sun.

Polarised light can be detected by many insects including ants, beetles, bumble bees, wasps, flies, certain spiders, caterpillars and also, curiously enough, by octopus, though its eyes are much like a mammal's which have not got this capability. Birds can also detect polarised light, notably pigeons. The path of a water-flea may be altered by shining down a vertical polarised light and then rotating it. A similar trick can be played on a horseshoe-crab, a prehistoric remnant unchanged for the last 250,000,000 years, the effect continuing down to a depth of about fifty feet.

With all these miraculous methods of orientation, let us pay tribute to the vulnerable baby turtle. Provided it is not caught by predators, it seldom fails to find the way from the egg to the broad ocean. It simply makes for that part of the sky horizon which looks lightest and this is nearly always to seaward. This is in contrast to the water-strider, whose mind is as polarised as its eyes so that, if alarmed, it always runs southwards across the pond.

STEERING BY THE EARTH

Certain primitive creatures, if unhappy with their surroundings, circle aimlessly until they find conditions more to their liking, after which they proceed roughly in a constant direction. At least this reverses them back to their original environment if they cannot find anything better. Naturally, for short periods, stabilizers will assist an animal to follow a straight line but only for short periods, owing to the limited memory.

Although animals are generally built symmetrically, they are not absolutely equal each side and tend to travel in circles. Human beings, moving over flat ground in pitch darkness, usually complete a circle in about half an hour, without much preference for the left or the right. As relatively few people are left-handed, the cause must be elsewhere and there is some correlation between differences in leg length and the direction and radius of circling. Perhaps this is one reason why elk travel across the tundra in long straight lines one after the other. All such animals are able to see behind them without turning the head and so the leader can make sure the route is not curving to either side.

Natural signposts. Steering is a matter of travelling in a constant direction. In polar regions, nature provides parallel ice ridges known as sastrugi which are formed by prevailing winds and are a feature of the great icecaps. At sea, prevailing winds produce parallel swells, notably in the Pacific Ocean. Also, as waves travel into shallow waters, friction slows them progressively and converts them into parallel waves running up the beaches and able to orient tidal animals. On land, winds affect the shapes of trees and the flower plumes of reeds and, in deserts, they carve out sand dunes of characteristic shapes and may also form parallel sand ridges.

On a more local basis topography, such as the general slope of the land, is a useful guide and the direction from which damp winds blow will encourage the formation of moss on certain sides of rocks or tree-trunks. A change in the direction of motion is suspected when the wind is felt at a different angle to the face or when clouds seem to scud across the sky on a different tack.

There are other more permanent but perhaps more subtle signs. The North American pilot weed and the prairie burdock align their leaves north and south. In southern Africa, the north pole plant leans towards the north. Termite nests run north and south and, in temperate latitudes, certain flowers bloom earlier when not shaded and may face towards the direction of greatest sunlight. A few temperate trees have the branches on their polewards sides raised to enable the leaves to catch more sun. Presumably animals can only learn about such signs if they already have some means of orienting themselves.

Compasses. A modern ship is steered by a gyro-compass, a stabilizer which detects the direction of the Earth's axis of rotation. It does this by measuring the vertical very precisely and remembering it for a long time. M. Picard's experiments with cockroaches, whose orientations were deflected by weights put beside them, hardly suggests gyro-compassing. For the modern gyroscope is not only far more sensitive than any animal sensors but also it has an inherent memory. Furthermore, it requires a knowledge of latitude and only works on a surface whose height is constant and so it is not used on land or in the air. Even at sea, it runs into trouble when course is continually being changed.

On the other hand, we find the magnetic sense is widespread among animals. The first discovery is said to have been made by a zoologist who opened the lid of a cardboard box containing a number of termite queens and found them all lying parallel, with their heads either east or west. He closed the lid and turned the box through a right angle but, next morning, they were again lying east or west. He put them in a steel box and they lay in all directions. He put a strong magnet on top of the box and they all turned at right angles to it.

Procedures of this sort have been tried successfully with many varieties of animals. When a flatworm or a freshwater snail has a magnet placed close by, it alters its heading. Whether it turns in order to null some lateral sensor, or whether it is lining itself up to the combined magnet and Earth's field, is not known. There is one argument in favour of the null. A slight deviation has a much bigger effect than quite a large discursion from a maximum signal, which is why the null is used for radio direction-finding.

Similar work on insects has revealed that a variety of flies, wasps, locusts and bugs can sense magnetism. Fluctuations in the Earth's field also affect the alignment of the dances of bees but this reacts the same way on the orientations of the bees that cluster round them. Sharks and rays among the fish also seem likely to be able to detect magnetism while, of the reptiles, the courses of turtles have been deflected by magnets attached to their shells.

Probably the clearest examples of the use of the Earth's magnetic field have emerged from studies of migrant birds. Dr. Gustav Kramer, the German zoologist, was able to prove that, at the time it should be migrating, a bird flutters in the direction in which it is going to travel, even if born and bred alone in a cage.[18] If the cage be rotated, the bird will realign itself in the original direction. The problem is, how does the bird know the direction?

Dr. F. W. Merkel of Frankfurt had some robins indoors and unable to see the sky. Unlike their British cousins, these little birds like to winter in Spain. At the time of their migration, they were all fluttering south-west. A steel cover was put round the cage to shield the birds from the Earth's magnetism, whereupon they fluttered up the bars in different directions. Finally, an artificial field of the same strength as that of the Earth's was generated within the cage and the robins then fluttered at an angle to this field which would have taken them to Spain had it been correctly aligned to the Earth's field. Conclusive evidence? Yes, but Dr. Hans Fromme failed to repeat it.

Similar experiments have shown that other birds, including gulls, have a magnetic sense. However, the discrepancies between Merkel and Fromme might perhaps have been resolved by the experiments of Wolfgang and Roswitha Wiltschko who showed that, if either the vertical or the horizontal field is artificially reversed, robins realign themselves in the opposite directions.[19] This is partly illustrated by Fig. 15(a) and (b), which shows the consequence of altering the vertical field. However, if both components are reversed, the birds return to their original alignments. Therefore these birds must rely on the slope or dip of the Earth's magnetic field and not on the direction in which the magnetic force acts. In electrical language, they detect the alignment of the Earth's magnetic field but not its polarity.

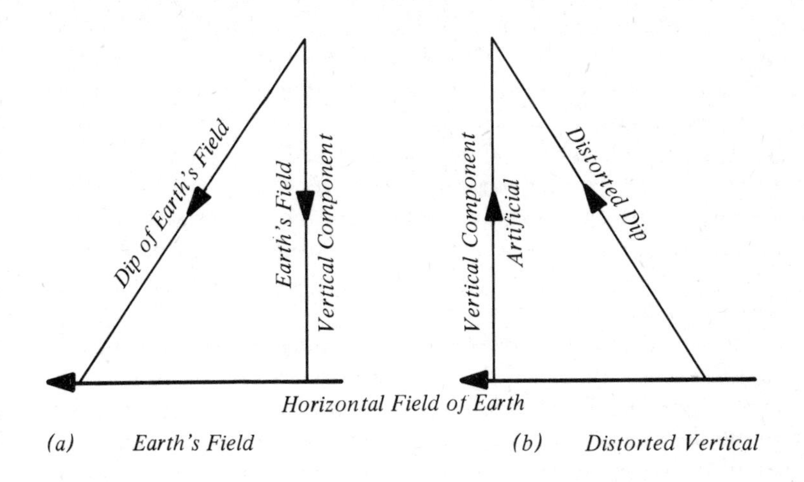

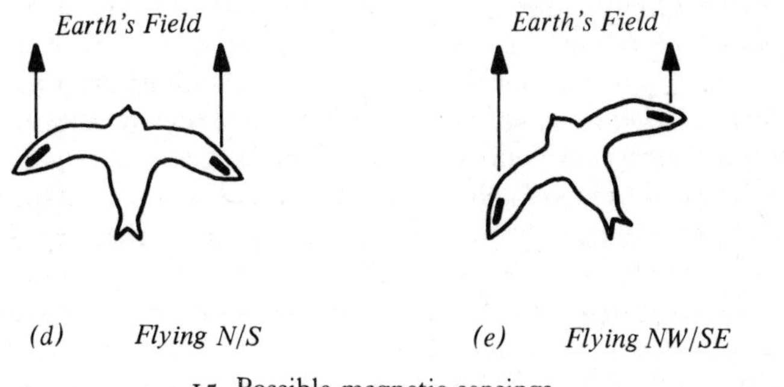

15. Possible magnetic sensings

In contrast to birds, relatively disappointing results have been achieved with mammals, though wild mice, bats and probably hedgehogs react to magnetic fields. Nevertheless it is difficult to explain certain feats otherwise. How can a seal dive down deep from its breathing hole and travel tens of miles under the pack-ice to the open sea? Also, strong magnetic fields affect the heart beats of monkeys, perhaps through their brains, and directly upset the brains of rabbits. Russian workers have suffered mental and psychological disorders when exposed to very strong magnetism and their hearts were also disturbed.

One would not be surprised to find human beings also had a magnetic sense, though it does not seem active enough to prevent them walking in circles in featureless country when it is pitch dark. Dr. R. R. Baker, of Manchester, arranged for some children, blindfold and fitted with apparently identical helmets, to be taken from their school along a winding route. Those wearing helmets with no disturbing magnets attached were able to point roughly towards the school. Those whose helmets were magnetised by a field directed towards the right ear, showed a tendency for errors 90° to the right and those with helmets magnetised towards the left ear, were deviated to the left.[20] Such results, duplicated by students on various occasions, suggest that alignment by the Earth's magnetic field is possible to human beings and that it is sensed by something in the head.

Magnetic sensors. Some bacteria in arctic or antarctic waters grow within their bodies magnets about a thousandth of a millimetre long composed of infinitesimal particles lying head to tail. Were the particles any smaller, normal temperatures would disturb the chain and, if any larger, the particles would form into separate magnet rings and cancel each other out. It is these minute magnetic sensors that enable the tiny lumps of jelly to determine which way is up, if they feel too cold, and which is down, if they are not cold enough. For, close to the poles, the Earth's magnetic field is nearly vertical and bacteria, being so small, live in a world of surface tension, not of gravity.

Insects carry magnetic substances in their abdomens. Certain fish, such as sharks, rays and the electric fish described in Chapter 5, are able to detect magnetic directions with sensors in the skins of their heads and their tails. It also seems likely that other lower animals detect magnetism with their bodies.

Pigeons are said to carry magnetic substances between their brains and their skulls. If *horizontal* magnets are attached to their heads, they become confused initially but soon learn to manage. Perhaps they find the direction of the applied magnetism does not change when they turn their heads, unlike that of the Earth's field, and so they are able to ignore the constant effect. However, when a reversed *vertical* magnetism is applied, pigeons go the opposite way. This shows they are detecting the dip of the Earth's field, just as the Wiltschko robins did.

Dr. M. J. M. Leask of Oxford has suggested that light might make part of the sensitive cells in the eye of a pigeon susceptible to magnetism, which would scatter the polarised light which this particular type of bird is able to detect.[21] The scattering might react on the retina in such a way that the direction of the Earth's magnetic field and its slope could be 'seen'. This brilliant suggestion solves very nicely a great variety of problems, but would render the magnetic sense ineffective at night.

It is known that, when pigeons are uncertain of the magnetic field around them, they flutter their wings, even when standing on the ground. This suggests they use their wings and, above all, the movements of these wings, which are greatest in the wing tips where the nerves run back at an angle to the fore-and-aft line of the body. Minute currents run along these nerves and could set up infinitesimal strains, but it is more likely that the movements across the Earth's magnetic field introduce minute electrical voltages in the conducting nerve filaments, and these voltages might be detected by the bird's brain.

Since a pigeon and a robin detect the dip of the Earth's field, it is important that the bird's wings travel up and down at an angle. They do this when they fly, as illustrated in Fig. 15(c), and also, wings move upwards and backwards and then forwards and downwards, an effect particularly noticeable when a bird is standing on the ground. Nevertheless, there is much evidence that magnetism is detected by the heads of birds and mammals and particularly by the brains of human beings.

Within the skulls of higher animals, countless minute currents of electricity flow in great profusion and would set up minute mechanical strains in the nerve channels due to the Earth's magnetism. We shall find this is one of the few plausible explanations for certain types of homing that will be discussed in the last part of this book. So the claims of some people that they sleep better with their head in a certain direction may be just as real as the responses of the termite queens!

Visualising the magnetic field. One interesting facet of Dr. Baker's experiments has been that the instinctive magnetic sensing by blindfold students has been largely discounted when vision was restored. This perhaps suggests what may be the great problem with magnetic orientation, namely, that it cannot be visualised, setting aside Dr. Leask's theory for the moment. An animal may point its head in the right direction but a sideways slope in the ground, or a cross-wind, will drift it at an angle. To correct this by using the vertical, it will be necessary to 'visualise' magnetic direction, by employing the feeling of magnetic alignment to fixate on a distant visual object, and to steer by that object for the time being.

At midday, the Sun would appear as a convenient object by which to 'visualise' the Earth's magnetic field and an animal could subsequently learn how the Sun's direction varies with the time of day. It is interesting to note that K. Schmidt-Koenig, of world-wide repute, has suggested the magnetic sense is instinctive but that the use of the Sun has to be learnt.[22] Once learnt, it seems much to be preferred as being more convenient. A pigeon may be flummoxed by magnetic anomalies in overcast conditions but, in clear weather, will be steering by the Sun and will discount them.

At night, the Moon might perform a similar visualising role. Alternatively, the sensing of the Earth's magnetic field might be linked to the direction in the night sky along which stars move neither up nor down, a direction which may itself be visualised by the Pole Star or the Southern Cross. The ability to translate a magnetic into a visual direction may be important if an animal wishes to travel at an angle to the field. This could be achieved magnetically if two sensors at right angles were available as in Fig. 15(d) and (e). To steer along the magnetic field, both sensors would have to be equalised but, to steer at 45°, one would be maximised and the other nulled. To steer at 90°, the two sensors would have to react equally but in opposite directions.

All these difficulties would be solved by Dr. Leask's suggestion, but this would require plenty of light. Also the ability to sense polarisation is not possessed by all creatures; no mammals have it. Naturally, an ability to visualise magnetic direction would lessen the need, in theory, to use Sun, Moon and stars. However, animals like to have back-up systems and so it may be assumed that not only can most animals detect the Earth's magnetic field but also that, by using it, they:

(a) Develop Sun, Moon and star compasses,
(b) Recalibrate these compasses when necessary and
(c) Retain the magnetic sense as a back-up in case the sky is overcast.

SPEED AND PRESSURE SENSORS

Animals travel at speeds which their forebears have found effective in past struggles for existence. Certainly self-sensors will tell a creature whether legs, wings or fins are making it travel faster or not and this may be important when a balance has to be struck between urgency and exhaustion. Nevertheless, it is not easy to see what would be gained by being able to record these speeds. The processes of evolution do not develop sensors which have no purpose.

Very primitive creatures may instinctively control their speeds according to whether they like their surroundings. A woodlouse has a skin that dries up quickly. If the air round it is hot and dry, it scuttles as fast as it can in a random direction until it comes upon somewhere cool and damp, such as the underside of a large stone or beneath a patch of vegetation. There it stays or moves about slowly, until hunger or the removal of its shelter sets it on its way again.

Speed sensors. A wide variety of creatures travel through the water, bees and birds fly through the air and land animals are affected by winds that are strong. So all creatures except the most elementary need to sense the pressure of air or water flowing over their bodies. Mammals detect air pressure with fur, whiskers and bare skin, birds with feathers, insects with hairs and, in particular with the two joints in their antennae next to their heads.

On the ground, such sensors act as wind velocity indicators, suggesting the direction from which the wind is coming and its strength. Thus the dung-beetle, somewhat unsteady on his long legs but equipped with sensitive antennae, puts his head down if the wind is coming from that direction, his tail down if it is aft and leans into a cross-wind. However, it is animals that fly which need to carry the most sensitive wind detectors.

The honey bee has two and a half thousand single eyes in each compound eye and there are little eyelashes in between which react immediately to air pressure. One important use of such sensors is to determine whether to fly or not, for strong breezes demand wasteful flying at angles to offset drift, and what may be gained on the swings of a tailwind will be more than lost on the roundabouts of a headwind, for the animal will then fly slower and be subjected to the adverse wind longer. A male moth, detecting the scent of a female wafted towards him, will not take off if the wind pressure on his antennae is greater than that which results from flying, knowing by instinct he will then be blown backwards away from his lady.

Blow-flies do not get airborne if winds are more then 8 miles an hour and other insects, including bluebottles, locusts, aphides and butterflies, have their own take-off regulations set by their antennae. On the other hand, some flying creatures find it difficult to get airborne unless they face into wind to obtain a little extra bonus in the form of initial lift.

It is important to appreciate that, once airborne, these sensors can no longer detect the wind. As shown in Fig. 16, a creature must be regarded as flying in a little block of air and the rate at which this block of air is moving cannot be detected by the air-pressure sensors. If the speed of the wind changes, the block of air and the animal inside will be accelerated so slightly or rotated so slowly that it is unlikely that stabilizers would detect it, though they would probably be well able to sense any gusts.

16. The 'block of air' concept

However, pressure sensors are able to help an animal fly in its block of air at speeds which give adequate lift. Dr. D. Burkhardt and Dr. G. Schneider of Würzburg have experimented with bluebottles. By pressing back their antennae, the insects could be made to raise their legs and change the patterns followed by their wing-tips. In particular, the further back the antennae were pressed, the faster the owners must have felt they were flying, for the slower they beat their wings, apparently believing their speeds through the air were too high.

Lateral lines. On the other hand, creatures that swim use pressure sensors for many purposes. Fish and amphibians such as newts and tadpoles carry very sensitive water-flow detectors on their bodies. When we eat a herring, we may notice a mark along its side suggesting a way to open it up with a knife. This is the lateral line. Each sensor in the line consists of a small bunch of hairs in a nodule of jelly not unlike those in acceleration and 'rotation' sensors.

Fish living in deep waters, where there is generally but little flow, have these sensors in the form of a row of pimples. In other fish, the lateral line may appear as a groove in the skin. The groove may be roofed over and filled with mucus, with holes through which the water pressure may reach the sensitive elements, thus developing into something that has been likened to an underground railway with frequent stations. The lateral lines join over the heads and jaws of fishes and a typical complex arrangement is shown in Fig. 17. When the fish is swimming, the sensors on the front of the head will record forwards pressure and the flow down each side will be sensed as reduced pressure. From this pattern, a fish will know how fast it is moving through the water.

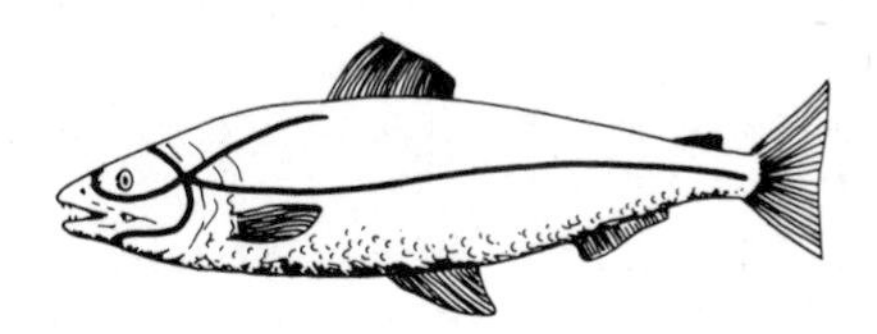

17. Lateral lines

The lateral lines are also able to pick up pressure waves reflected from objects close by each side, and this makes it possible for the fish to tell the speed at which these objects are passing by or are being passed. Thus, from the rate at which headway is being made compared to the rocks each side, the fish can sense the rate at which the stream is flowing. This will be useful if the fish needs to find a slow-running spot in order to rest. The ability may be even more important to a male salmon battling its way upriver. To have found the best way through the turbulent waters may be crucial when the final pathetic battles between exhausted fish take place in the spawning grounds.

Pressure. The pressures that indicate speed are in addition to the steady pressure of the water itself, which increases with depth. Thus lateral lines help a fish to keep at a certain level and this acts as a back-up system to the swim-bladder. Some fish can even detect changes in the atmospheric pressure. The loach, a fish with a retiring nature, used to be kept in aquariums before the days of meteorological forecasts by radio. Normally it hid in the sand with only its eyes peeping out but, twenty-four hours before a storm, the drop in air pressure would cause the fish to swim excitedly up and down the glass sides of its tank. Another 'weather fish', the gurnard, grunts when pressure falls suddenly, whereupon prudent Mediterranean fishermen tend to return to port. Finally, an honourable mention for a species of prawn which can sense its depth to within a few centimetres. On its skin are cells covered by liquid and with an outer membrane. The liquid is electrically charged, the current building up a thin layer of compressible gas. This gas balances the pressure of the water and any imbalance leads to an increase or decrease of electricity producing more or less gas to restore the balance. These electrical changes pass to the prawn's brain and tell it to go down or up.

SUMMARY

1. *Stabilizers*. Control motion and assist steering, including interceptions.
2. *Steering.* May be assisted by natural features but is mainly dependent on
   (a) *Astronomical* means using
      (i) *Sun*. Widely used, relying on direct or on polarised light and combined with time. Bees direct their fellows by waggle dances.

5. *Fiddler crab*. The skin of this crab has an inbuilt clock: it changes colour at high tide to camouflage itself; moved to an aquarium, the crab will continue for a while to be quite a reliable tide-predictor. *Stephen Dalton, Natural History Photographic Agency*

6. *Turtles*. They lay their eggs in the sand. The newly-hatched turtles look for the sea where the sky is lightest. *Spencer Arnold, BBC Hulton Picture Library*

7. *The cheetah* uses proportional navigation when pursuing its fleeing prey: it keeps the direction of its quarry as constant as possible by using its change of rotation sensors. *Brian Rogers, Biofotos*

8. *Reindeer*. Perhaps they travel across the snowfields in long straight lines one after the other to avoid travelling in a circle: the leader can see behind to ensure the route is not curving to either side. *Heather Angel*

(ii) *Moon*. Probably by estimating highest point in sky.

(iii) *Stars*. Using general rotation patterns assisted by stars that rotate very little.

(b) *Magnetic compasses*. May be universal. Some animals only measure slope of Earth's field. Inconvenient because visualisation difficult so magnetic sensors may be used to

(i) *Develop* astronomical compasses,

(ii) *Recalibrate* these systems if necessary, and act as a

(iii) *Back-up* if sky is overcast.

3. *Speed*

(a) *Land and flying animals* detect winds when on the ground.

(b) *Fish and amphibians* have lateral lines which measure speed through water and also rate of flow.

4. *Pressure*. Speed sensors measure pressure and may also indicate depth in water.

PART TWO

Senses for Hunting and Avoiding

We have to be relatively close to an explosion to feel the shock, for most of the energy is used up blowing things to pieces. However we hear the bang a long way away because sound waves shift the air through which they pass by very small amounts. Creatures that hunt or wish to avoid being hunted will obviously start from a distance and so we would expect them to employ waves.

We know about two types of wave, electro-magnetic which includes light, infra-red and radio, and mechanical waves such as vibration and sound. However, there are other sources of information which operate from a distance. Scent diffuses through water and air while electricity is conducted by impurities in water. So this part of the book has been divided into three chapters:

(a) Electro-magnetic waves,
(b) Mechanical waves and
(c) Chemical and electrical systems.

Although we expect senses that direct motion from place to place to operate from a distance, we find that contact systems often use the same sensors and it is difficult to decide where one ends and the other begins. Thus heat is allied to infra-red, touch merges into vibration and taste into smell.

3

Electro-Magnetic Waves

The longest electro-magnetic waves are those produced by radio and radar transmitters. There is no valid reason why an animal that can detect magnetism should not sense these waves. However radio waves from outer space are greatly attenuated as they pass through the atmosphere and man-made radio aids certainly do not floodlight the surface of the Earth. It would therefore be necessary for animals to produce their own radio waves and either pick up the reflections or use the transmissions to deter predators or to attract prey and mates. Quite apart from the formidable electrical problems, had these transmissions been possible they would surely have already been picked up and monitored.

Thus we shall be interested only in the shorter infra-red, visible light and ultra-violet waves, though there is evidence that some insects sense X-rays and these waves also affect the blood of rodents, such as mice, and probably other creatures. We shall examine visible light and ultra-violet first because they are generally picked up by eyes. We shall then turn to heat and to the infra-red radiations it produces.

VISION

Eyes. Generally, the whole bodies of single-celled creatures are affected and, from this, specialised eyespots have evolved, particularly on the backs of primitive animals to keep them upright. On some transparent organisms, the eyespots are underneath, with a lightproof patch on top to give Sun direction by shadow.

From such beginnings, single-celled eyes, illustrated in a notional form in Fig. 18(a), have appeared on many simple creatures including clams and caterpillars. Some of these eyes are made to vibrate so that

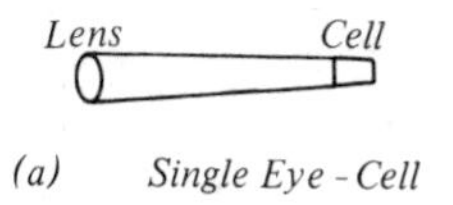

(a) Single Eye - Cell

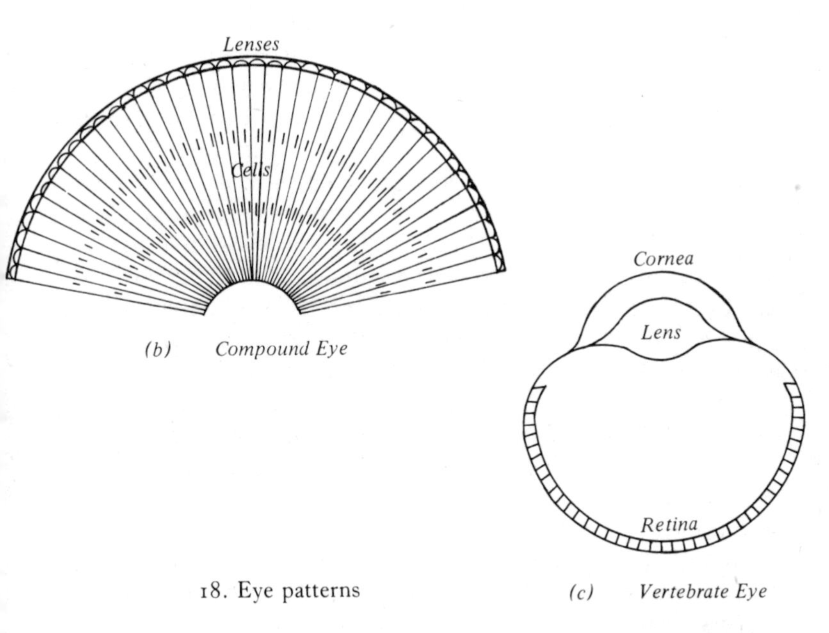

(b) Compound Eye

(c) Vertebrate Eye

18. Eye patterns

they can collect more information. Worms carry eye-cells on various parts of their bodies. A timid variety of eel, hiding from the hungry by day, has an eye-cell in the tail to advise it when darkness has fallen and it is less dangerous to venture forth. Insects usually have three on the tops of their tiny heads and a spider, living mostly by touch, may have six or eight which can be moved in any direction. Bees seem to use single eye-cells to tell them when to set out in the morning, so as to start gathering nectar at first light, and when to fly home to clock-in before it gets too dark.

Fish carry simple light cells, trout and pike having particularly sensitive exposure meters. Amphibians may use them for orientation and reptiles need them to tell when to change colour for camouflage. Dr. Maurice Burton has suggested the eye-cells in some lizards may inhibit motion, keeping the creatures lying in the morning sunshine until it gets too hot, whereupon the sunbathers rise up and leave the area. Even in the human being, a vestigial 'pineal' eye is embedded in the brain and has links with the other eyes, but apparently no navigational significance.

Fireflies and other insects that operate at night have complicated eyes, in which a hemisphere composed of a number of very special lenses steers what light there is across a gap, to be concentrated onto an inner hemisphere of sensitive cells. Some creatures living in the black ocean deeps have similar eyes, but with reflectors instead of lenses, and these are also used by shrimps and lobsters.

However, as a general rule, insects and crustaceans have a number of single eyes combined to form a pattern and making up a 'compound' eye, as shown in Fig. 18(b). The worker ant has, each side, six such eyes combined to form a minute raspberry. These have developed in large crustaceans and insects into wonderful compound eyes which reach a climax in the dragonfly which has 28,000 each side, adding up to around the number of words in this book.

Vertebrates and those cunning submariners, the octopus and the squid, have combined the multiple lenses of the compound eye into a single lens. The eyeball is filled with liquid and the sensitive cells, packed into the back like a camera film, are known as the retina. The arrangement is illustrated in Fig. 18(c). Light from a distant object reaches the eye in almost parallel lines and is bent by the lens to focus at a point on the retina. However, light from a near-by object diverges and has to be bent by a greater amount. Fish may move the lens forward to give the light more room to converge but most vertebrates squash the lens by muscles so as to increase its curvature and its power to bend light.

A transparent shield or cornea protects the outside of a vertebrate's eye and there is liquid between it and the lens. Parallel rays coming through the air from a distant object and striking the eye are therefore given a first bending by the cornea before the second bending by the lens focuses the light onto the retina, and this is shown in diagrammatic form in Fig. 19(a). On the other hand, if underwater, there will be liquid outside the cornea and this initial bending will not occur. All the work has to be done by the lens itself which therefore needs increased curvature as shown in Fig. 19(b). Because the rounded cornea would contribute nothing in the water but would increase water resistance and be more liable to damage, fish have developed flat corneal coverings.

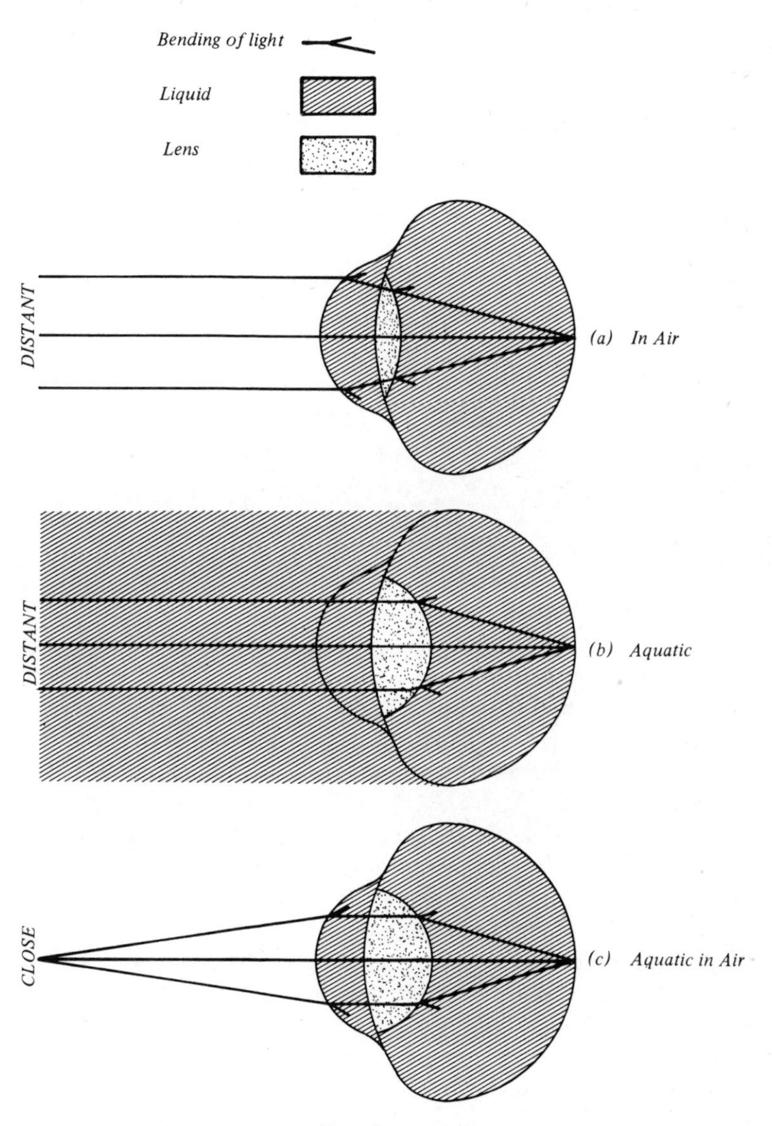

19. Vertebrate vision

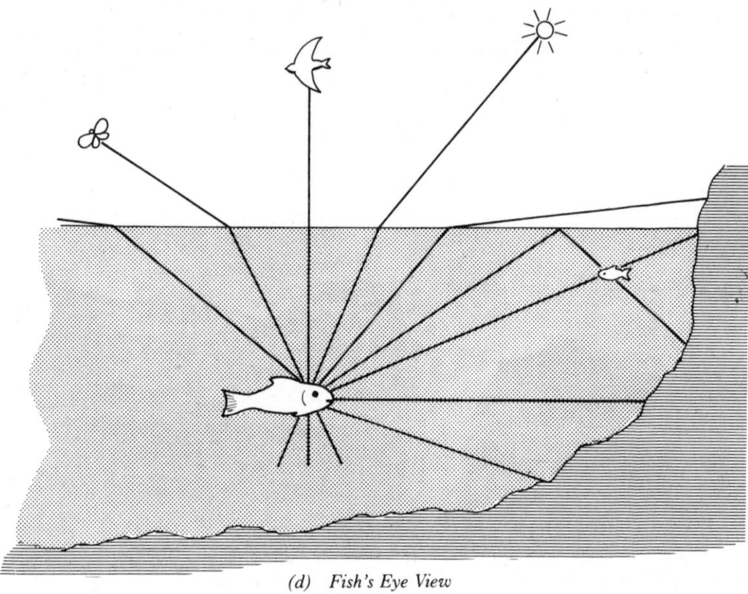

(d) Fish's Eye View

Penguins live by catching fish in the sea and, to see prey at a distance underwater, their lenses also need the increased curvature. However, when they come out of water, the situation arises shown in Fig. 19(c). All the lines of light in the eye are the same as in Fig. 19(b), but the bulging cornea with the liquid inside also bends the rays and, unless they emanate from an object close by, they cannot be focused on the retina. As a result, the penguin on land is very short-sighted and, until a human approaches to within a yard or so, the bird is apt to take him for just another penguin.

Other aquatic animals have similar problems and, when whales pop their heads out to look at a boat, they too are short-sighted. They do not do very well in water either, for they are believed to find it difficult to focus their distant tails. To see underwater, some birds such as cormorant, auks and diving ducks, have very soft lenses whose shapes they can alter considerably or else, like terns and gannets, they sight their prey from the air and strike before it has time to get away.

Fish have other problems. When light from above meets the surface of the water, it is bent downwards, the bending increasing as the light strikes the water at finer angles until, when almost parallel, it is bent by 45°. Therefore, although a fish can see in all directions, its view of the world above the water is squashed into a 45° cone and this is shown in Fig. 19(d). The rest of the upwards vision outside the cone comes from reflections off the underside of the water surface so that, as suggested by the small fish in the figure, the same things may be seen twice. When the water surface is ruffled, light bendings and reflections become hopelessly confused which is why inquisitive whales come up to peep.

As a result, the hippopotamus and the crocodile have eyes raised above the tops of their heads so that they may look out when their bodies are awash. The four-eyed fish of South America has the best of both worlds, one pair being below the surface and one pair above. The archer fish, which shoots down insects flying just above the water by squirting from below, does not attempt to solve the problems of bending light but waits until the target is directly overhead. The water bullet spreads out into droplets like gunshot and it usually manages to score a hit though one pet archer fish, with a screwy mouth which put his shots off centre, never hit anything and never learned how to.

Fish that live deep in the oceans, where the Sun's rays hardly penetrate, need very large eyes even to see other fish as shadows above them and the Gigantura carry what are virtually telescopes. Land animals that work a night shift likewise have to make the best use of what light there is. They tend to have very wide lenses fitted to eyes so big there may be no room for the muscles that should turn them. Thus an owl has to be able to rotate its head through more than 180° each way and a toad's eyes have filled his skull to such an extent he has to use them for swallowing, so that he blinks as he gulps.

The retina. In a vertebrate, the light-sensitive retina carries two types of cell illustrated in Fig. 20(a), namely

(a) Cones, which are separate sensors and also detect colour, and
(b) Rods, which are cells linked together.

Cones are packed closely and can present a fine-grained picture. People, monkeys, birds, lizards and fish have, in each eye, a concentrated area of cones known as a fovea. A human being has, in each square millimetre of fovea, no less than 125,000 cones, more than twice the total number of words in this book, making it possible to distinguish differences in direction of a tenth of a degree.

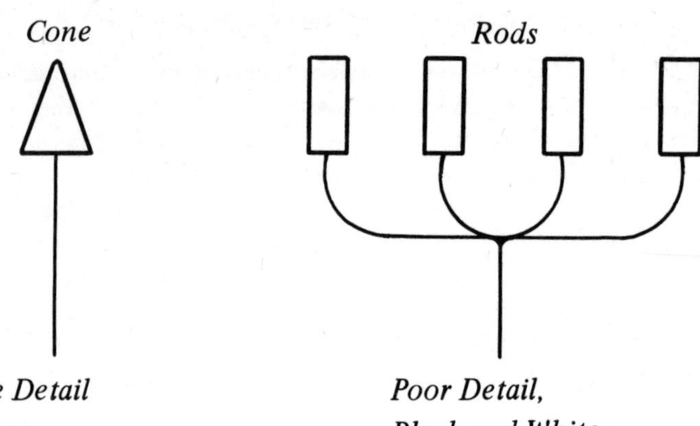

(a) Types of Eye-Cell

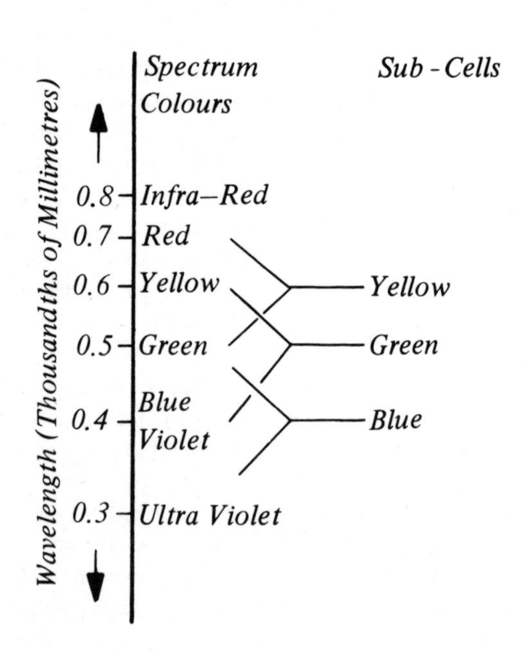

(b) Colour Vision

20. The retina

A buzzard has eight times as many cones as we have in each square millimetre of fovea, giving it a vision comparable to a human being with a good pair of binoculars. Swifts and swallows have two foveae in each eye, one looking sideways and the other concentrated ahead to pick up the minute insects on which the birds feed as they fly. One may see a swift strike at a butterfly which subsequently flutters gently down to the ground but, in that lightning dart, the bird will have taken its head and its abdomen.

As we can see from a rainbow, the cones of the human eye distinguish a range of colours from red, the longest wavelength, through yellow, green and blue to violet, the shortest. Longer than the red waves are the infra-red, which extend up to about three tenths of a millimetre and gradually become very short radio waves. At the other end of the spectrum, the violet waves become ultra-violet which stretch down to one hundred thousandth of a millimetre and then merge into X-rays.

The spectrum is shown in Fig. 20(b) together with the three types of sub-cells within the cones of the human retina. These record colour in terms of maximum signals at yellow, green and blue[23] but spreading out to the colours either side. When we see red, only the yellow sub-cells are activated, while yellow gives a strong reaction to the yellow sub-cells and also affects the green ones and so on down to violet light which affects only the blue sub-cells.

The spectrum is shown in wavelengths to suggest why, the smaller the creature, the less likely it is to be sensitive to the longer red waves but the more likely to detect the shorter ultra-violet waves. Thus a bee can detect yellow, green, blue and ultra-violet. Dr. K. von Frisch has shown it only sees intermediate shades between its four colours as grey and red and so it may only distinguish a dozen colours, compared to the 250 which may be separated by a human being such as an interior decorator or a fashion designer.

Unlike the cones, the linked rods do not distinguish colour but only black and white. They are connected together in a way that enables them to record movements, a feature very significant to a hunter or to an animal that is being hunted. Also, by teamwork, they respond to very faint light. Human beings have 15,000 rods combined with 6,000 cones in each pin's head of retina outside the fovea. Small fish which may need to see in the faint light deep down in the ocean may have as many as 25 million rods per square millimetre. An owl, able to see by

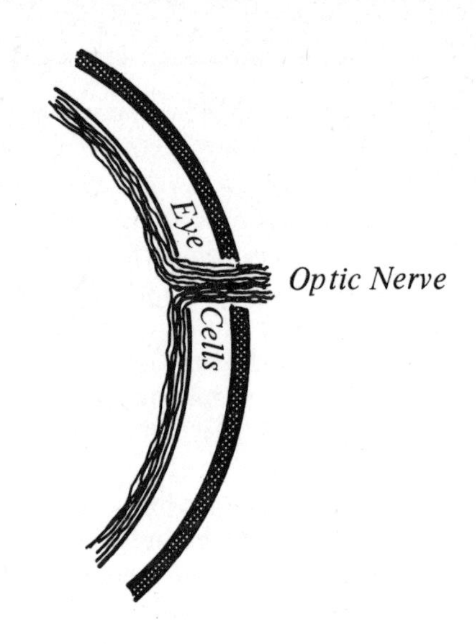

(a) The Retina

21. Retina and blind spot

the light of a candle a hundred yards away, has its rods linked together in hundreds. However, the picture painted by the rods is naturally far more coarse-grained than one produced by cones.

Although the cells are so closely packed together, there are still gaps between them and, at the back of the eyeball, there may be a material which reflects any light and gives the rods and cones a second chance to use it. This produces the 'eye-shine' of animals adapted to see in faint light, from which the reflectors in road surfaces have become known as 'cats'-eyes'. Whales and seals have 'eye-shine', to help them see in deep dark waters, and so do mammals that hunt at night, but humans, apes and pigs are not so favoured by nature.

The retina is a part of the brain which has budded off from it at a very early stage in the development of the embryo. As a result, it consists of a complex of nerves with light-sensitive cells added almost as an afterthought on the wrong side, so they are partly shielded as shown in Fig. 21(a). Where the nerves run together before leading out from the eyeball, they block the visual cells completely and form a blind-spot on the retina. If the left eye is closed and the right eye looks at the cross in Fig. 21(b), provided the book is held level and about a hand's breadth away from the eye, the black blob will disappear. You

(b) The Blind Spot

may have to move the book a little closer or further away and it is necessary to concentrate on the cross. Because people are not aware of this blank area, the brain must be filling it in.

As a part of the brain, the retina may do certain things on its own. For example, it would take too long to signal to the brain details of an object moving at high speed across the field of vision and for the brain then to analyse the information and pass instructions to the eyeball muscles. The retina therefore does the job on its own, but passes on to the brain what it sees. It is therefore not surprising that the nerves leading from the retina to the brain generally contain less channels than there are sensitive cells.

Within the eye of a bird, there is a heavily pigmented area connected with the retina which is known as the pecten, and it is believed to be able to absorb the intense light from the Sun. Certainly it has a flow of blood which might be able to carry away the heat generated. It therefore seems likely a bird will align its eye so that the Sun shines on the pecten and this would automatically give to its brain the direction of the Sun compared to the vertical. The movement of the eye necessary to keep the Sun aligned might help the bird to keep a track of the motion of the Sun across the sky. As we would expect, the pecten is very small in night owls, but large in eagles and hawks.

The brain. We have already seen how the brain fills in the blank spot on the retina. It also sees what it expects to see. If an oval shape is drawn on a piece of paper and memorised and then a round plate is tilted until it appears to reproduce the same shape, on placing the paper beside it, the plate is seen to have been tilted too far. The brain is aware the plate is round and makes it look rounder than it really is, so that we tilt it to an extent greater than we should. Thus when we expect to see somebody, we see them and then the brain realises the pattern is not quite right and the result is the well known 'double-take'.

The brain also tends to imagine that it can see the colours picked up by the edges of the retina, which contain mostly rods that only detect black and white. Thus we can 'see' the colour of a car that goes past us when it comes from ahead because the brain knows what colour it is. But, if it comes from behind, it will not appear coloured until it has travelled well into our field of view.

Not only does the brain paint its own picture but also, assisted by the retina in higher animals, it may produce results from quite unpromising visual signals. Dr. Anton Hajos of Innsbruck fitted spectacles to himself and some students which so distorted their vision that things which were straight looked curved, objects that were fixed leapt up and down when the eyes moved, and coloured fringes appeared where there were none before.[24] After a week, the wearers were again seeing things as they knew them to be.

When the spectacles were removed, confusion reigned once more. Lines were curved but in the opposite directions, objects leapt down and up and colour fringes reappeared. Fortunately it did not take long to readapt. In a similar exercise, inverting lenses were tried because the retina 'sees' things upside down and left to right like a camera film. The victims again managed to adjust, but readjusted rather faster back to normal vision.

One must expect the power to make the best of unpromising material to be possessed also by animals. For example, each separate single eye in a compound eye may be sensitive to light coming in from an angle as wide as 30°. Nevertheless, it can be shown that the same insect can detect two points of light only half a degree apart. This ability must be the result of a sorting either by the brain or by the networks of nerves joining the outputs of the individual single eyes. In addition, many insects scan with their eyes, moving them from side to side and up and down to improve their vision.

By using similar methods, the human eye does better than it ought to. The eyeball muscles make it tremble continuously, which has been proved by fitting small reflectors onto contact lenses.[25] As a result, the eye scans across whatever it is looking at and fills in any gaps between the cells in the retina. Were it not for this, our eyes would probably see things in a pattern rather like a newspaper picture and certainly we would not be able to read the small print. The brain, or the retina, allows for the trembling so that we do not see it.

This trembling has another interesting consequence. If it is inhibited by optical means and, at the same time, any large movement of the eyeball is counterbalanced, the brain gets used to the steady response from each eye-cell and ceases to take any notice of it! The object goes grey and eventually black. If the object moves, the pattern changes and we see it in silhouette as if against a blackboard. The eyes of many animals do not tremble and it is possible they too only see things that move or they have to move their eyes to see things that are standing still.

The retina of a frog is comparable to that of a human being but the human brain, with its remarkable flexibility, can cope with an extremely wide range of information. By comparison, Professor J. Y. Lettvin, of Cambridge USA, suggests a frog 'sees' only four things.[26] Two do not spell danger. They are a straight edge which may be the side of a pond or a blade of grass, and a convex shape which the toe-in of the eyes recognises is not beyond grabbing distance. This of course is the characteristic shape of an insect to be eaten but a leaf, or even a small frog, may be stuffed into the frog's mouth. The other two items are a change of contrast, which could be a stork approaching, and a rapid darkening suggesting something descending from above. In both instances, the reaction of the frog is to leap towards anything blue, hopefully a pond but equally well a sheet of blue cloth.[27]

An ant, shown a black and white flower shape or a dot, will go for the latter which suggests the entrance to its hill. A bee looking for nectar will choose the flower shape but, if on the way home, it will opt for a horizontal black line. The hoverfly, needing stems to climb up and get a better take-off, likes thin vertical lines, whereas caterpillars that are clumsy and tend to fall out of trees go for thick vertical lines like tree trunks and will climb a human leg. The water-beetle prefers horizontal lines, like the bank to which it hurries for shelter from hungry fish. Dr. Rudolf Janders of Freiburg, who undertook many such remarkable tests, suggests the insect brain has not the capacity to analyse everything it sees and has to learn to react only to patterns likely to be of most use to it.

Yet we should not underestimate the brains of lower animals, particularly insects. Their central nervous systems will have built-in patterns so that a bee will sting if if feels it may be squashed, but these automatic reflexes may be modified by memory. For example, a solitary wasp will drag prey back to its hole according to an imprint of surroundings gathered on the way out. Nor should we forget that quite primitive animals appear to have emotions. An ant that is lost stands rooted to the spot. A bee trapped behind a window-pane will die, not from exhaustion or lack of food, but from what appears to be panic. Yet, if a bird should take off its abdomen, it will continue unperturbed to eat honey, which will pour out of its thorax unconsumed, until eventually the creature is no longer alive.

HUNTING AND AVOIDING BY SIGHT

We have to look at hunters at the same time as the hunted not only because even the vegetarians need to hunt for mates but also because so many animals who prey on others are themselves hunted by larger predators. This is particularly noticeable at sea, where big fish eat little fish and little fish eat lesser fish. So it goes down to the smallest krill and these curiously enough are also eaten by the largest aquatic animals, the huge baleen whales.

Nocturnal mammals, birds, reptiles and amphibians, and also fish that live in deep oceans, have ample rods in their retinas so that they can see in poor light, but their sight is not good and they often use other senses. Vertebrates that hunt by day have rods all round the sides of their retinas to pick up any movements of prey over as wide angles as possible. They then turn their eyes towards the movement to find out what it is, using the cones in the centres of their retinas. The pattern is illustrated diagrammatically in Fig. 22(a)i.

Vertebrates that are hunted need rods all over their retinas to detect movements which may spell dangers. They cannot afford to concentrate on only one movement, for many predators hunt in packs. To identify enemies, they also need cones in all directions, as caricatured in Fig. 22(a)ii. In practice, all animals sometimes require concentrated vision and so they tend to have more cones in the centres of their retinas, even if they do not have true foveae.

Hunters and the hunted also differ in the placing of the eyes, as suggested by Fig. 22(b). The former concentrate on one quarry at a time and have to be able to judge distance very accurately for the final strike, particularly the falcon stooping at 100 miles an hour. Therefore, the eyes are at the front of the head to give wide binocular vision forwards, but the mounting of the eyes close together tends to restrict the field of view of each of them. On the other hand, a creature

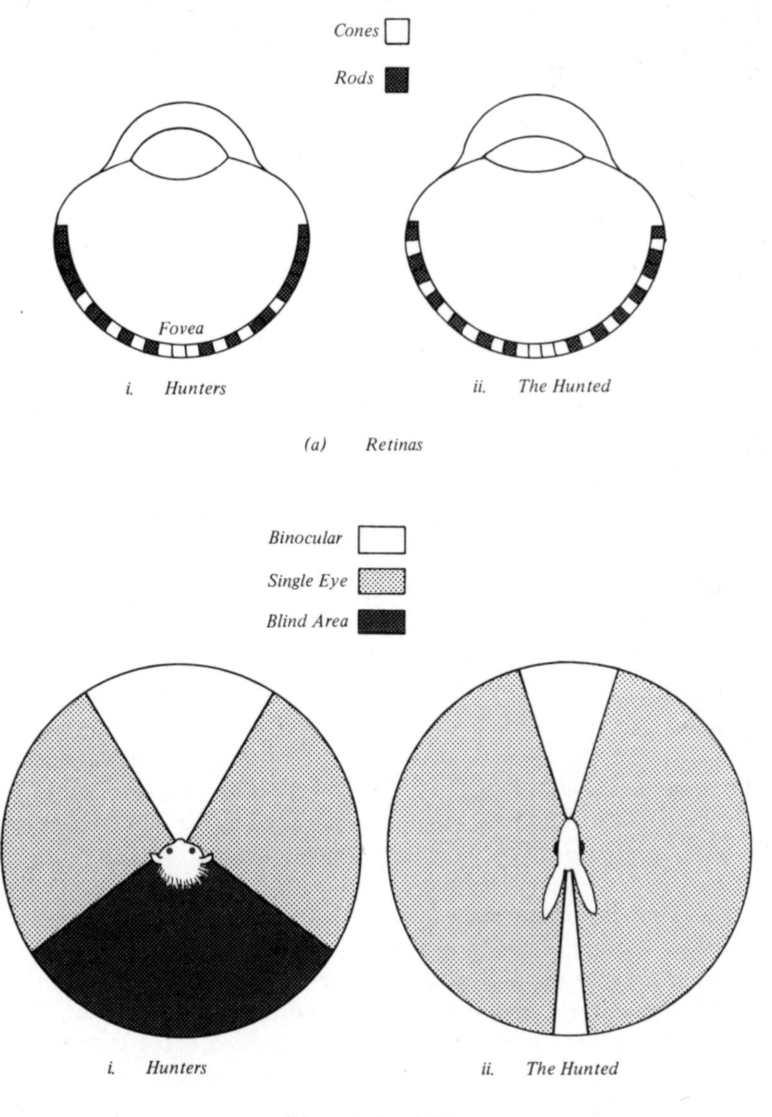

22. Eyes of hunters and hunted

that is hunted needs to be able to see all round. Its eyes will therefore be at the sides of its head where each will have the maximum field of view and there will be some binocular vision forwards and also, in order to judge the distance of a pursuer, aft as well.

Apes and monkeys, though not hunters, need good binocular vision forwards because, as they swing through the trees, they cannot afford to miss a hand-hold. Human beings also concentrate on vision forwards to enable them to manipulate objects with their hands and to use tools. Let us quote a few examples. Man has a field of view of 160° ahead of which 120° is stereoscopic. So each eye has a field of view of 140°. A dog may have a rather better field of 180°–190° in each eye, giving a total field of 280° and 90° of binocular vision.

The individual eyes of mammals that are hunted and of birds, except for birds of prey, have the widest fields of view of all. For example, each eye of a hare covers 220°, the pair giving an all round view with 30° binocular vision ahead and 10° aft. A whale, on the other hand, covers only 125° each side because eyes that project or look forwards will not take kindly to being driven through salt water with silt in suspension. Dolphin, however, must surely have stereoscopic vision ahead, otherwise they could not judge distances when they leap out of the water through hoops, a favourite trick in oceanariums. Fish generally have less problems because they do not generally travel so fast and, of course, their corneas are flat.

Because human beings rely so implicitly on their sight, it should not be assumed that the same applies to animals. Many use sight for feeding and for the final stages of a capture but, for hunting, other senses are frequently paramount. A lion, sighting its quarry a mile away, may then stalk it head-down by scent, knowing instinctively that to put its head up and take a look will cause its prey to take flight.

Many hunters chase prey visually, in which event the latter may have a trick to play. Deer and rabbit have white patches behind, wonderful targets on which pursuers can concentrate without really noticing the animal in front. When all seems lost, the quarry will suddenly turn sideways and the hunter will lose the distinctive target for a brief moment, often enough for the terrified prey to increase its safety margin, at least temporarily. By this means an antelope, which can only travel at about forty miles an hour, may escape from the cheetah which can reach seventy miles an hour but only for a very short burst.

Camouflage. It may be questioned whether camouflage has a navigational significance. It may have in human terms. Thus, in World War II, dummy towns were built, complete with searchlights and anti-aircraft defences, in order to mislead bomber navigators. Furthermore, we accepted in the Introduction that avoiding, the negative of hunting, is a navigational activity and camouflage is an avoidance system based on reflecting light waves of specific frequencies, that is, of certain colours. We shall also find that a few animals seek to deceive hunters by producing sounds and vibrations of selected frequencies, a form of sonic camouflage.

Animals, whether preying on others or being preyed upon, try to disguise their outlines by light and dark patches that break up their silhouettes, and one may walk past a young speckled fledgling without seeing it. Woodland creatures wear spots to merge with the sunlight filtering through the trees, desert hunters are sand coloured and arctic creatures have white coats, others changing their colours into white for the winter if they live on the edge of a snowline. Chameleons and stick insects can alter their colouring to suit their backgrounds. Certain crabs cover themselves with algae and weed and, if in an aquarium with pieces of sponge rubber on the floor, they will stick these onto their shells using saliva. Some small river fish have a simpler solution; they stir up the mud and escape behind the cloud.

Mammals, including whales and dolphin, together with birds and fish, have light-coloured undersides to their bodies to balance shadow. There are exceptions, such as caterpillars that crawl along the undersides of plant stems and the Nile catfish with a dark belly and a light back, but it happens to swim upside down. However, the greatest weapon of the hunted is to avoid exciting the rods of the hunter by keeping absolutely still. At long range, this may not be too difficult but, if the hunter is close, the breathing of a mammal or the twitching of an insect's legs may give it away. It is interesting to note that certain snakes and lizards will not take prey unless they see it moving.

Camouflage also hinders the finding of a mate. Mammals that hunt, and those that eat grass and are exposed to attack, have ample rods to detect movements and so they identify mates by shape and size rather than by colour. Birds on the other hand may be brightly coloured, particularly on their breasts, where they do not give too much away to hawks above but yet attract mates. Fish are similarly brightly coloured, though the skins on their backs generally reflect blue light which makes them more difficult to see in the water. Butterflies are brightly coloured in flight but, when settling, they close their wings and show drab undersides.

Although concealment is the norm, many animals that sting make themselves quite obvious to predators. Once stung, the hunters learn to avoid the danger colouring of red, orange or yellow with black bands, which appears on wasps and hornets, on poisonous South American frogs and on the bellies of many dangerous snakes. Creatures that attack dangerous snakes, unless like the mongoose they are experts, seldom live to learn greater wisdom but it seems most animals, as well as people, have an inborn fear of these reptiles.

Naturally, harmless creatures copy these retaliators. The hoverfly assumes the exact colouring of a wasp. Harmless snakes develop red undersides with black stripes. The hawk-moth larva even imitates a snake and many butterflies have patterns like eyes on their wings which remind small hunting mammals of an owlet. Other insects have false heads on their abdomens so that, when molested, they fly off in an unexpected direction. Lizards flatten their bodies, puff out their throats and turn sideways to make themselves look bigger, and may open their mouths wide to show a brilliant warning orange inside.

Other animals wear unmistakable uniforms because they are unpleasant rather than dangerous, so that an animal that eats one will leave others alone, the sacrifice of one being economical in terms of protecting the others. The skunk, of course, has distinctive black and white markings but it is not generally known that the brilliant kingfisher tastes nasty. The monarch butterfly, which feeds on the poisonous milkweed, is coloured in an arresting fashion. So the edible tiger-moth has bright yellow and orange rear wings which it displays hopefully when it is apprehensive.

Luminescence. Although messages are carried along nerves by electrical currents, the power to move muscles is transported about the bodies of animals by compound molecules of a substance known as ATP, which releases energy when broken down. It happens that certain nocturnal insects and creatures living in the dark depths of the oceans are able to break down ATP not to move muscle but to produce light,[28] used generally to attract a mate or to lure prey. These illuminations are produced quite cold and represent almost 100 per cent conversion of energy into light, compared with 4 per cent for an electric lamp and 10 per cent for a fluorescent tube.

Male fireflies produce 1/40 candlepower flashes for around a third of a second every five or six seconds and the females reply exactly two seconds later, thus keeping to a precise code. Sometimes, swarms of males light up together in trees, producing startling effects as if a magic wand had been waved across the foliage. One type of female firefly has learnt the code of another and, when she has attracted the male, she eats him, retaining her love only for the male of her own species. Glow-worms and will-o'-the-wisps also illuminate themselves and some beetles flash regularly, like little lighthouses, or switch double headlamps on and off. The 'railroad' worm has little green lamps along each side and a red spotlight on the head.

Striking light displays occur deep in the oceans where sunlight cannot reach. At 30 feet, 90 per cent of the light is lost and at 120 feet, 99 per cent has been absorbed. Below three or four hundred feet, the depth associated with continental shelves, there is virtually no light at all and the creatures that live in these depths have to provide their own. So squids carry proper searchlights, with lenses and shutters, to blind prey and dazzle enemies.

Minute male creatures carry lamps to attract females and discourage rivals and the females respond with friendly flickerings. Other fish, rather larger, look for these lights to devour their owners. Still bigger fish carry illuminations in their mouths to welcome little hunters in. Viper-fish may have hundreds of lights in their mouths and small fish, with other creatures, swim happily in to be consumed among the luminescence.

Sea-dragons, with fearsome coloured jaws, carry twenty port-holes each side of their bodies and trail a coloured anchor fore and aft.[29] The angler fish has a fishing rod growing out of its back, with a luminous worm acting as bait dangling in front of its mouth. The bait is withdrawn smartly if prey threatens to nip it off and the offender is then sucked into the ferocious jaws. Another creature has a rod four times as long as his body and so, when he attracts prey, he has to pounce on it.

Sometimes illuminations are used for defence. Certain shrimps let out a cloud of sparks when frightened. Other fish carry phosphorescent mites in little bags under their eyes, with openings to permit free entry and exit to their passengers. When threatened, they eject the squatters and hope the glittering cloud will confound their enemies. A few deep sea fish glimmer faintly underneath to avoid casting shadows from the faint sunlight onto predators below. Predators however have one great advantage over their prey. They may stock up their luminescence by devouring other luminaries.

It is important to stress that, despite all this exotic variety of creatures in ocean deeps, life is extremely sparse. Angler fish are built so that they can swallow creatures bigger than themselves and therefore a meal, though it may be rare, can at least last a long time. The female has such a poor chance of meeting a male that, when she does, she allows him to act as a parasite on her body in order to keep him permanently for herself.

WHAT CREATURES SEE

Although it has 15 million sensitive eye-cells, the human eye, as an instrument, would surely be a disgrace to any competent optician. The lens is so poor we have to cut out the edges by screwing up our eyes if we want to look at anything really small. However, the cells react to the whole visible spectrum from red to violet but not to ultra-violet nor infra-red. As daylight fades, the extreme ends of the spectrum weaken first and the other colours dissolve into black and white as vision is taken over by the rods. Apes and monkeys have visual powers similar to ours, though some of the smallest may be deficient at the red end of the spectrum.

Mammals on the whole are regarded as being short-sighted and, because most of them have a large number of rods in their retinas, their colour discrimination is suspect. Indeed, it may be the brightness of the rag rather than its redness that excites the bull. Giraffe confuse orange, yellow and green, and so may believe their camouflage is better than it is, but hedgehogs react very strongly to yellow. Horses, sheep, pigs and squirrels appear to have poor colour discrimination, while the golden hamster is said to be quite colour-blind.

Of the vegetarians, the rabbit has a retina comparable to that of a human being while the horse has the largest eyes of any mammal, but its ability to focus sharply is probably affected by the ramps in its retina, which permit objects at various distances to be focused simultaneously. Other hoofed animals have good eyesight, particularly deer, but elephant and rhinoceros see little further than ten yards from their noses.

The cat family have extremely good vision and so do those remorseless hunters the killer whales. Most bats and shrews seem to see little, the exceptions being the bats that eat fruit and those that catch fish, detecting them partly by the ripples they cause and gaffing them with long back claws. The vision of badgers, voles, mice and rats seems to be limited to a distance of about ten yards, probably adequate for their purposes, while moles, living on worms and insects underground, have little use for eyes.

The finest vision is certainly to be found in birds. Expressions such as 'eagle-eyed' and 'eyes like a hawk' show a long-standing recognition of the extraordinary visual acuity of birds of prey and vultures. A kestrel will stoop from two hundred feet to pick up a small vole half hidden in the grass. Even tiny humming birds have better vision than human beings. Pigeons have better retinas than we have and a jay will see a gnat thirty feet away. Certainly the aquatic penguin is short-sighted on land but not underwater and the only bird with poor sight is the kiwi; until predators were introduced into New Zealand by the settlers, it had no enemies and needed neither wings nor long sight.

The vision of nocturnal birds, including owls, is blurred by the extensive linking of rods in their retinas but nevertheless the large eyes of barn owls allow them to catch flying insects and crickets at night. The farmyard cock wakens us early because he can see in light ten times as weak as his human owners. Also, from their plumage, we can infer that birds recognise colours. Gull chicks, from the moment they hatch, beg for food by pecking at red spots on the beaks of their parents and will peck also at red spots on inanimate objects. Birds also recognise shapes, for even a newly hatched chick will peck instinctively at anything shaped like a grain. We may therefore anticipate that birds will be adept at identifying objects around them.

Of the reptiles, snakes are short-sighted but can apparently discriminate finely at short range. Lizards will turn towards small insects many feet away. Crocodiles and alligators certainly see very well and even a simple slow-worm can identify moving creatures at a distance measured in yards. However, it seems likely that reptiles are largely colour blind, which might account for the sombre nature of their appearance, but there are some exceptions, lizards apparently sensing green, yellow and red, but not blue. Amphibians also appear to be myopic but this may be due to limitations imposed by their simple minds on things seen to be outside grabbing range and not apparently dangerous.

Except for those that live in muddy waters, fish have good eyesight, as anglers well know. Sharks in particular have acute vision and, when they get hungry, they will frighten fish out of their homes in the coral reefs and into open water, there to be chased and eaten. We may infer that fish have colour vision for they are often brightly coloured, particularly on their backs. As we go deeper into the sea, red light disappears first and so ocean fish tend to have poor red vision but can see ultra-violet. Indeed, many of them and also crustaceans and the ubiquitous octopus can detect polarised light. On the other hand, freshwater fish, whose operating depths are naturally limited, cover the normal range of human vision.

Insects are short-sighted and only recognise objects which are close. A bee needs to be near to a bloom to distinguish it from a leaf and an ant will only notice another which is less than an inch away. However, a dragonfly, patrolling up and down in a woodland glade, will dart off ten or fifteen feet sideways to capture a midge. Insects that collect pollen from flowers have good colour discrimination but, except for butterflies, tend to be deficient at the red end of the spectrum, making up for this by being able to detect ultra-violet and to sense the polarisation of the Sun's light. Sometimes, colours are used without discrimination. The greenfly starts with a liking for blue so that it flies up into the sky when the weather is warm. It is then drifted about for a time, after which it develops a yen for yellow, the colour of young shoots on plants, and descends to be captured by astute gardeners using bright yellow bowls filled with water.

Simple creatures generally have limited vision and, for example, a caterpillar with six eyespots each side will manage perfectly well if eleven of them are obscured, which suggests they are only exposure meters. A worm will not react at all if lit by a red lamp. In tidal waters, clams and other shellfish see with parts of their bodies which are exposed when they open their shells, but these react mainly to light and shade. Jellyfish have eyespots in their skirts and float amongst the plankton which includes simple animals that flail themselves along and curl up if a shadow descends from above, even if the shadow is from one of their own limbs.

Although sight may not be the most universally useful of all the senses, it has been helpful to consider it first because it illustrates nicely the importance of the brain. In vertebrates, the retina is a highly specialised computer which completes certain elements of sorting and even directs some instinctive actions. In other senses, special lobes of the brain handle the nerve impulses. Thus the sensitivity of a creature to a stimulus may often be inferred from the size and complexity of the lobes devoted to breaking down the signals into their distinctive parts and combining these parts together to build up the necessary impressions.

INFRA-RED AND TEMPERATURE

Infra-red. Anything warm gives off heat or infra-red radiations. It is not always possible to determine whether it is the radiations that are being detected or the consequent warming of the air. A female mosquito will detect, with its antennae, the small change of temperature from an area of exposed skin at a distance of ten feet. A bed-bug climbs onto the ceiling and wanders about until it feels that something below is warmer than a sheet or blanket, whereupon it drops and feasts.

Rattlesnakes, pit-vipers and moccasin snakes carry each side of the head, between snout and eyes, small pits about 6 millimetres deep and three millimetres across as shown in Fig. 23.[30] At the bottoms of these small pits are concentrated five times as many heat receptors as on the whole of a human body, so that a difference in radiation due to a temperature change of a tenth of a degree can be sensed at several feet.

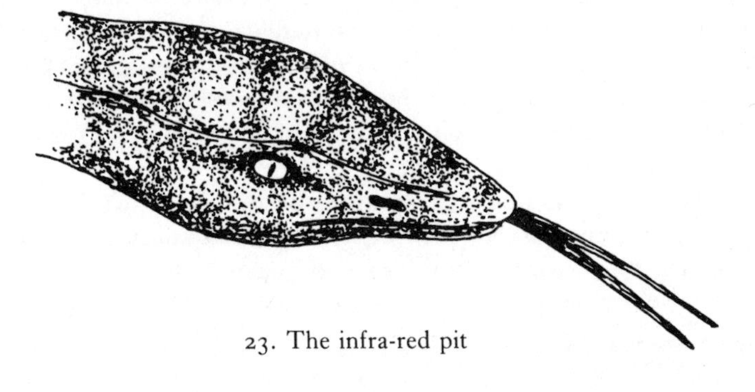

23. The infra-red pit

Each pit is designed to receive only a very accurate cone of radiation and these cones overlap along a line dead ahead of the snake.

In pitch darkness, these snakes will pounce unerringly on a small warm electric light bulb wrapped in paper which has no scent. On the other hand, if the pits are covered, the creature will not be able to capture a live mouse in the dark. By moving the head up and down and from side to side, the snake can detect the shape of the heat picked up and discover whether it is a rat that can be eaten or a mongoose that might kill. The pits do not react to the overall temperature of the surroundings nor to gradual heat changes but only to the abrupt alterations of temperature caused by the infra-red outlines of the prey, due for example to temperature differences of as little as ½°C. Thus a pit-viper will detect anything warmer than its surroundings and it is difficult to suggest how prey can disguise this difference.

Temperature. Warm-blooded mammals and birds are able to control their temperatures by thermostats, generally in the brain, and by sensors in skin, mouth and nose which are of two types, one reacting mainly between 20°C and 30°C and the other peaking between 40°C and 45°C. By comparing these outputs with the thermostat, the brain can decide whether its surroundings are too hot or too cold and whether it needs to lose or gain heat to maintain the correct temperature, which is 37°C in the instance of mammals and 40°C for birds.

If mammals feel cold, they try to warm themselves up by muscle activity which, if the surroundings are very chilly, may take the form of shivering. If too warm, they cool themselves by evaporation either by sweating, which is most efficient in relatively hairless humans, or by using the dampness inside them, which causes the panting of cats and dogs or, in the instance of small rodents, by licking themselves. At the other end of the scale, the elephant will spray water on his ears, his skin elsewhere being too thick to pass on the cooling effect to his body. By such means, a polar bear will live happily on a Greenland ice-cap or in a Californian zoo.

Birds also raise their temperatures by muscle activity and lower them by panting but their control tends to be less effective than that of mammals. In extremely cold weather, warm-blooded animals may become dormant, which involves lowering the metabolic rate, the rate of living. Indeed many small animals become dormant in the autumn in anticipation of the severe winter. Detection of temperature has other uses. When in the spring it rises above 10°C, horseshoe-bats sally forth in search of dung-beetles.

The female night moth may hot herself up temporarily to attract a male but, in general terms, all creatures other than mammals and birds take their temperatures mainly from their surroundings and are said to be cold-blooded. They usually try to live in conditions between 10°C and 40°C but there are many exceptions. The desert ant is only happy at about 50°C while the mosquito infests the Arctic tundra and even the ice-caps of the north and one small beetle, which is able to produce its own anti-freeze, can survive for long periods at −40°C.

Reptiles maintain temperatures between 25°C and 35°C by moving to warmer or cooler places and they die quickly if they become too hot. In strong sunlight, a snake will adjust matters by the amount its body is in or out of its cool burrow. Amphibians may use lateral lines for heat measurement but their responses are less easy to predict. A frog will not jump out of water which is heated up slowly even when the temperature has risen to fatal limits.

Fish, surrounded by water from which they cannot escape, are naturally extremely sensitive to its temperature for to them it is a matter of life or death and sensors are distributed all over their bodies. Salt-water varieties may detect a change of as little as 1/30°C, or a similar difference between their heads and tails. Freshwater fish are generally less sensitive but can feel differences of 1/10°C. However, it is worth noting that there is a fish that lives quite happily in Labrador fjords in water at −2°C.

Insects carry heat sensors in their antennae but some, including those in the caterpillar stage, also use their mouths. If placed on a surface which is warm at one end and cold at the other, an insect will generally position itself at a point which is consistent with an ability to detect temperature with an accuracy of ±1°C. Social varieties of bees, wasps and also termites maintain their communal homes with a constancy of this order, the honey bee keeping its brood chamber at 34°C. Lower animals are progressively less affected by hot or cold conditions and can often recover from extreme desiccation or freezing.

Parasites that live on mammals and birds are, however, very sensitive to the temperatures of their hosts which explains why cat fleas do not stray onto human beings. Chicken mites, creatures a millimetre long, leave their hosts by day because the birds become too hot but return to the warmth at night, even if it happens to be represented by an electric clock. The dreaded tsetse fly stays in sunlight when the temperature is less than 30°C because cattle will be grazing, but it seeks the shadow when it becomes hotter and its victims tend to collect under trees. Creatures such as these carry thermometers in peculiar places. The sheep tick has them in its front legs.

SUMMARY

1. *Primitive creatures* may have single eyes, used also by higher animals as exposure meters.
2. *Crustaceans and insects* have compound eyes.
3. *Vertebrates* have single lenses and retinas. In particular:
   (a) *Hunters* have binocular vision ahead with retinas having colour sensitive cones in their centres and, at their sides, rods that detect movements in black and white.
   (b) *Hunted* animals have all round vision with rods and cones in all directions.
4. *Vision* depends on how much the brain contributes. Also
   (a) *Underwater animals* see land distorted to a 45° cone.
   (b) *Birds* have excellent sight and colour discrimination and the ability to recognise objects.
   (c) *Mammals* are generally short-sighted.
5. *Camouflage*. Used by hunters and hunted, the latter sometimes disguising themselves as dangerous or unpleasant.
6. *Infra-red*. Certain snakes carry detectors for hunting warm-blooded prey.
7. *Temperature*
   (a) *Warm-blooded*. Mammals and birds control their temperatures.
   (b) *Cold-blooded*. All other animals take temperatures largely from their surroundings.

4
Mechanical Waves

We have discussed the significance of waves and the most obvious mechanical examples are those that occur in the sea, due to vertical and lateral oscillations of the water itself. However, the senses in which we shall be interested detect alternate compressions and expansions of particles of air, water and even land which are shunted to and fro like clattering trucks in a railway siding, the shuntings travelling at speeds of 1,100 feet per second through elastic air, or around 5,000 feet per second through relatively incompressible water, though the compressions and expansions are extremely small. Any distinction as to whether the results are vibrations or sounds depends on whether they are felt or heard, though sometimes the same sensors are used for both. The sensors that feel vibrations may also detect the pressures and relaxings of touch, which may perhaps be regarded as extremely slow vibrations.

VIBRATION AND TOUCH

Touch. The sense is used in the final struggle between hunter and hunted, but this is not a part of navigation. However, it will be useful to mention the example of the solitary wasp which digs a hole and lays its egg in it and then goes off to collect fresh living food for the emerging larva to eat. It therefore has to paralyse a creature without killing it by stinging it at a carefully chosen spot and so it is not surprising that it can cope with only one type of prey. The wasp then drags the paralysed body to the hole, drops it in and covers it over.

Hairs are, as we have seen, very efficient tactile sensors. Side whiskers enable a cat to creep towards its prey without rustling against the vegetation either side. The walrus uses them when searching for clams in the sand. But the most sensitive of all animals is probably the mole, with pimples on snout and tail because it may travel forwards or backwards down its burrow. All these are examples of the use of the sense for hunting, but certain caterpillars have irritant hairs to discourage predators.

There is also one spectacular tactile navigator, the jerboa, a little long-tailed creature and a great jumper that only goes about at night. Two bristles, each as long as its body and depending from its nose, act as blind landing aids to enable it to touch down safely and it steers itself during leaps by trailing its long tail over the ground.

A few animals even extend the sense of touch by using tools. The woodpecker finch will poke a stick or a cactus spine into crevices in the bark of a tree to flush out insects. A chimpanzee will insert a stick into the entrance of a termite nest and enjoy the defenders that cluster round the offending weapon and grip it.

Vibrations. A whirligig-beetle lays its antennae on the surface of the water to detect prey which is struggling. In the same way, the backswimmer in a pond feeds on creatures that fall into the water by mistake, detecting vibrations between 5 and 500 per second and moving towards the disturbance until it can see what is struggling. Arrow worms catch small creatures swimming past by detecting vibrations up to 20 cycles per second, but their effectiveness is sorely reduced if the bristles on their heads are trimmed. Anemones likewise capture organisms and even the brainless jellyfish will move the fringes of its skirts towards a tiny crustacean which is disturbing the water as it struggles along. But perhaps the most remarkable of such hunters are the immature larvae of flies who have learnt to live in crude petroleum around oil-wells. They feast on the less adaptable and peculiarly careless creatures who fall in by mistake.

A spider appears to sense vibrations by its feet and it is very sensitive to the period, distinguishing between a large and dangerous creature and a small fly that can safely be enmeshed and eaten. Normally it hides to one side of the web but runs to the centre when it detects a promising tremor. A male spider, wishing to attract the attention of a female but wary of being eaten, will twang her web and, if she feels encouraging, she will twang back in reply.

In the marine world, vibrations may well be the most important source of information and will be detected by lateral lines which may be used for:

(a) *Hunting.* A small perch will snap at a glass tube the thickness of a pin's head provided the tube is moved even by a very small amount. At the other end of the scale, sharks gather from far and wide to vibrations that suggest a creature in difficulty, for this arouses them to ferocious attack. They will hurry towards a fish struggling in a net but if, when they arrive, it is released, they will ignore it unless they happen to be hungry at the time. Deep sea fish, whose eyes can see little in the darkness, rely heavily on lateral lines. They have them in their illuminated beards and affixed to the lighted bait they carry on long fishing lines. This is just as well, for they cannot afford to lose these valuable portions of their anatomies.

(b) *Spawning.* A male will flap his tail gently at a female to persuade her to lay, or he will waggle furiously at a rival who decides to join in the fun. This tail wagging is an integral part of mating. It is so stimulating a factor that a female salmon may be persuaded to spawn simply by stirring the water with an oar.

(c) *Avoiding.* Just as fish can detect a predator, so the lateral lines will enable them to travel in shoals. If one of their number be attacked, it will flee and the others close by will sense this and scatter, affecting the remainder until the whole shoal disperses, hopefully confusing the hunter.

We see that the lateral lines are a most important part of the armoury of a fish. In the ocean depths, they are certainly much more important than sight, and this is probably true even in shallow sunlit waters.

Reptiles also sense vibrations and a snake lays its lower jaw on the earth to detect things moving about. Birds carry sensitive vibration sensors in their legs to detect a predator climbing up the tree trunk towards them at night. A squirrel, high up in a tree, will react instantly to drummings of a certain type on the trunk, having an instinctive fear of climbing martens, though it may never have seen one and these predators may have long disappeared from the locality.

The mole detects the movements of an earthworm and burrows towards it. The vibrations alarm the other worms and they struggle up out of the ground, to wriggle away as fast as they can. Even a human being may place an ear to the earth to detect a distant footfall.

Just as vibrations emerge imperceptibly from touch, so they develop gradually into sounds as their frequencies increase. It is therefore not surprising that hearing sensors are similar to those in lateral lines.

HEARING AND SOUND

Fig. 24 gives a rough overall picture of the frequencies normally heard by animals. As a general rule, oscillations of less than twenty per second will be sensed as vibrations. At the other end of the scale, oscillations become faster and notes higher until, once again, the ears cannot cope with them. Although a child may hear up to 30,000 cycles per second, the usual upper limit for an adult human being is 20,000. Above that, vibrations can only be sensed through the bone of the skull and are no longer heard by the ears. Accordingly, they are known loosely as 'ultrasonic' waves although, as we see from Fig. 24, many animals can use them.

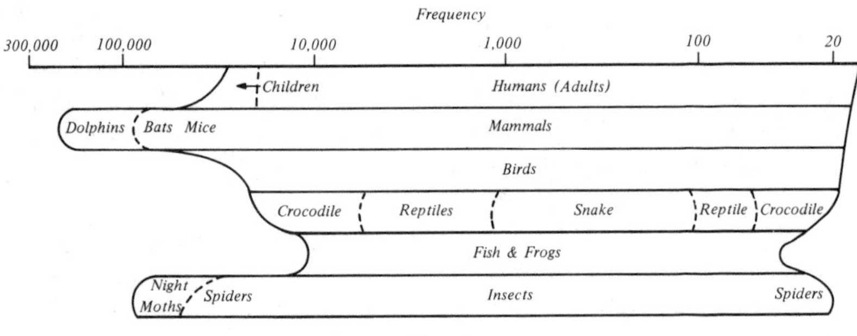

24. Hearing

Hearing. A few flying insects hear with their antennae. Those attached to the male mosquito vibrate between 450 and 600 cycles per second which coincides with the wing-beats of the female, but the males' wings beat faster and so they do not pursue each other. If blobs are stuck onto the ends of these antennae, it alters the rate at which they vibrate and the male can no longer detect the female. But this is only one of a variety of insect ears.

A more typical insect sound detector is an eardrum, with sensors round the edge but open to the air on both sides, so that sound can be detected coming from either direction. Sensors of this type can hear

9. *Elephant seals* are the first creatures man has recognised as using local 'dialects' distinguishing the two halves of a colony. *Brian Hawkes, Natural History Photographic Agency*

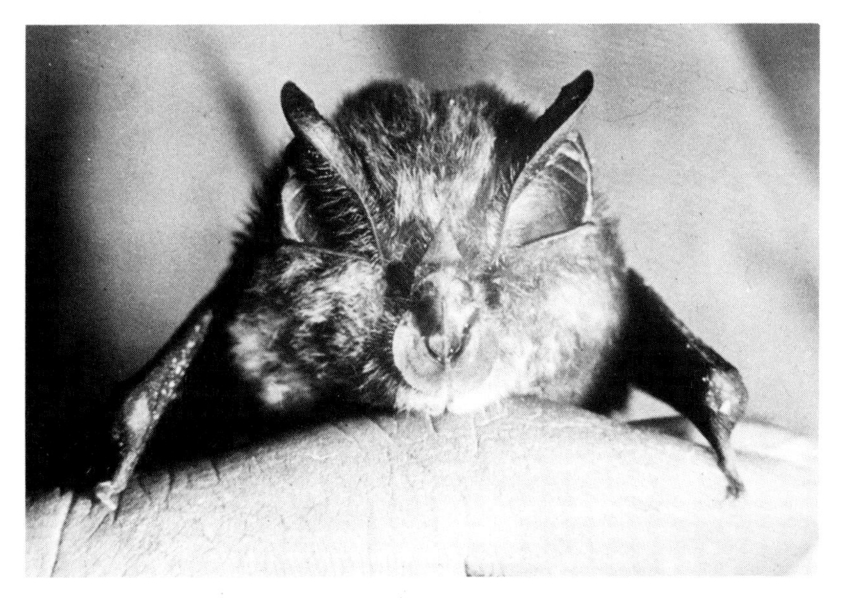

10. *The horseshoe bat* has large reflectors that confine sound waves roughly within a cone of about twenty degrees each way and, incidentally, make it look fearfully ugly. *I. R. and L. A. Beames, Natural History Photographic Agency*

11. *Manx shearwaters*. These little birds use the Sun's vertical and their internal clocks to navigate, which enabled one of them to home on its nest on the Welsh coast twelve days after leaving Massachusetts on a three-thousand mile ocean crossing. *D. A. Smith, Aquila Photographics*

frequencies normally from 250 to 10,000 cycles per second. Grasshoppers carry these ears on their front knees and on their abdomens close to their waists, while crickets have them on front knees and tails.[31] Cockroaches also use tails but cicadas have ears on their abdomens close to their waists.

A wide range of hearing is enjoyed by spiders using ears on the last joint but one of all their legs. The smaller the spider, the higher the note to which it responds, thus ensuring that the high buzzing of a small insect caught in the web is immediately detected. The larger spiders can tackle insects with a slower, more powerful wing beat, though this does not always mean much for the little hoverfly buzzes exactly like a wasp. Very high frequencies are sensed by night moths which have ears close to their waists and generally just behind the wings. They can sense sounds up to 150,000 cycles per second, the frequency used by the bats that hunt them.

The bodies of fish are composed of liquids whose density is so close to that of the water in which they live that sound travels straight through the fleshy parts, but reappears at the junctions with their gas-filled swim-bladders. Their ears, which are generally developed from sandwich-sensors, are therefore embedded inside between swim-bladders and brains. By such means, fish react particularly to low tones, rays hearing bass notes perhaps as a uniformly low sound irrespective of frequency.

Many freshwater fish, including carp, catfish, minnows and loaches, have the swim-bladder linked to the ears by small pieces of bone which have budded off the vertebrae and are so arranged as to magnify the sound vibrations. This gives greatly improved hearing with a sound range generally from 16 to 7,000 cycles per second but sometimes going up to 16,000 cycles per second. Above these frequencies fish may only hear high notes as a constant tone without much discrimination.

As we saw in Chapter 2, these bone linkages help a fish to detect changes of water pressure. But their real significance lies in the way they have been developed into the hearing systems of reptiles, birds and mammals, including human beings. Ears of these higher animals consist of eardrums which, unlike those of insects, are open to the air on only one side. On the other side, a bone linkage amplifies the sound and passes it to the sensors which detect its frequency. These sensors form integral parts of the labyrinths which stabilize such animals.

Amphibians have exposed eardrums and are able to hear but some react to sounds in peculiar ways. Shout at a frog and it will ignore you. Touch it on the back and it jumps. Shout and touch it at the same time and it positively leaps. Reptiles have similar eardrums except the burrowers and the snakes, the extent to which the latter can hear being arguable. Reptilian ears show signs of being able to discriminate between tones, crocodiles in particular having a wide range and turtles doing well on low notes.

Birds have ear-holes which, except for the ostrich's, are concealed by feathers. Their hearing organs are well developed and they pick up sounds on eardrums open to the air on one side and pass them through an amplifying linkage to the hearing sensors. However, mammals probably have the best hearing of all with ears developed even further than a bird's and with large external ear-flaps to deflect the sounds into the ear-holes, though these have disappeared in the aquatic whales and dolphin.

Horses, elephants and dogs can follow verbal instructions and so they must be able to hear the range of tones used by human beings to communicate. In addition, mammals generally hear in the region known to man as ultrasonic, hence the Galton whistle to summon a dog. Cats have similar hearing, going up to 50,000 cycles per second, an exception being the pure white domestic varieties which are apt to be deaf.

Small rodents, such as mice, can hear even higher sounds and we would expect this. The higher the frequency or rate at which sounds strike the ear the shorter must be the intervals between the crests of the waves. Just as small insects detect shorter light waves than larger animals, so the ears of smaller animals are better attuned to short sound waves. Bats can hear well above 100,000 cycles per second, notes which can also be detected by the night moths on which they prey. However, the toothed whales and their smaller cousins the dolphins go twice as high as this and can hear frequencies higher than any other animals.

Sounds. We shall now consider the sounds that animals make. In general they do this to attract mates. Thus fruit-flies sing and beetles click. Male grasshoppers and crickets rub legs and wings together, which seems to please the females much more than their personal appearances. The cricket sends out a morse signal at 300 cycles per second but the note seems to be immaterial for the lady responds to the morse. Cicadas have drum-like organs that chitter and many beetles make quite loud noises, the death-watch beetle summoning ghosts in derelict mansions by banging its head against hard objects.

Animals in the sea appear also to use sound when searching for a mate rather than when hunting prey. Sea-horses click and sea-mares click back. The bosun fish whistles and his lady pipes up in reply but, if a rival appears, the two will grunt at each other until one is discomfited and departs. Anemone fish also fight but the battle ends when one acknowledges defeat by quacking like a duck. Indeed, it has only recently been appreciated how noisy fish can be. Even shrimps and prawns add to the din.

Some fish drum and croak using ligaments inside, with the swim-bladder acting as a sounding board. The sun-fish grates his back teeth while others grind their front ones. As described in Chapter 2, the gurnard grunts, while the toad-fish roars every half minute and others make chugging sounds six times a second, all against a background of growls, hisses, groans and snores. Indeed, ocean fish are so noisy they excite sonobuoys listening for nuclear submarines and detonate acoustic mines lying in wait for battle cruisers.

Frogs and toads make sounds ranging from a deep croak to a happy tinkle. Many reptiles also make noises of a kind. Crocodiles hiss, bellow and roar, and snakes hiss and rattle. Some harmless snakes have even learnt to vibrate their tails among dry leaves and produce the exact sound of a rattlesnake.

Birds have developed sound for what might be called social purposes. These include claiming territory and attracting a mate, which produces the well known songs of cock birds in the spring. In addition, birds recognise each other by sound. The male emperor penguin leaves his wife in the rookery holding the baby and departs for many days to collect fish. When he returns, as short-sighted as ever, he recognises her call among those of hundreds of other birds. Many birds, including starlings, jays and mocking-birds, can even reproduce the calls of others, and parrots, budgerigars and mynah birds can mimic human beings, a capability credited to no mammals.

Mammals, like birds, make sounds to express anger, fear and excitement or to attract mates and they listen also for warning sounds such as rabbits give by drumming on the ground. Those which hunt in packs give tongue when on to prey and they do this to help each other, even though they may dispute fiercely over the carcase.

Whales and dolphin communicate with others of their kind and this is known because they make certain sounds only in company and never when alone. The noises of white whales, on the other hand, which have given them the name 'sea canaries', are probably due to bubblings from their blow-holes. It is interesting to note that elephant-seals, huge creatures weighing up to four tons, are the first creatures which man has recognised as using local 'dialects', which may separate one half of a colony from the other when breeding off the Californian coast. Fig. 25 summarises the uses animals make of sounds and also introduces the hearing systems they use for direction finding.

ANIMAL TYPE	HEARING SYSTEM	USE OF SOUND	DIRECTION-FINDING
Mammal	Ear flaps*, eardrum & labyrinth	Commun-ication	Ear flaps* & time-differencing
Bird	Eardrum & labyrinth	Social purposes	Time-differencing
Fish	Bladder & labyrinth	Mating	—
Insect	Eardrum, antennae, etc.	Mating	Sound shadow

*Not in whales, dolphins or certain seals

25. Systems involving sounds

HUNTING AND AVOIDING

The ichneumon fly, with ears in its feet, detects the sound of a larva inches deep in a tree, the larva moving about to listen to reverberations which warn it not to burrow into someone else's living quarters. When the larva hears the drilling of the fly, it stops moving at once and so may avoid the awful fate of acting as a living larder. However, most animals use their hearing to measure the direction of prey or predators at a distance.

Direction-finding. Insects, with hearing organs on various parts of their bodies, probably rely mainly on shadow, the masking of sound by other parts of their anatomy, and this is particularly helpful if their ears are sited just behind the wings. Grasshoppers and crickets with ears in their front legs can walk towards noises with little mental effort, detecting directions by changes of strength in the signals reaching the two knees. Tree-crickets raise their wings to reflect sound. If an enemy should approach, they lower their wings and the sound seems to recede or, by turning and twisting the wings, they can beam the sound to either side, thereby producing a ventriloquial effect.

Fish must find it even more difficult to find direction, for their hearing organs are inside their bodies and very close together. From a few directions, fish may be able to use shadow but not many organs will produce this for, in general, they will have the same consistency as water. Also, because sound travels five times as fast through water as through air, it is much more difficult to find the difference in arrival times of a sound at the two hearing organs. Probably the only way they can get any impression of direction is by using their lateral lines.

An ear separation of say one inch and sound waves coming directly from one side represents a time difference of about $\frac{1}{13,000}$th of a second in air, sound travelling at about 1,100 feet per second. In water, the time would only be $\frac{1}{60,000}$th of a second, as the sound moves at about 5,000 feet per second. If the sound were from ahead or behind, the waves would arrive simultaneously in both ears. Unfortunately, the waves will arrive almost together over an appreciable angle. The animal has therefore to turn its head slightly from side to side, to find the mid-point between the two directions along which the difference in time arrival is just perceptible.

The brain uses quite a simple system to measure differences in times of arrival of sound waves.[32] A signal from a nerve can either elicit a response or inhibit it. The nerves from the ears are cross-connected so that the one which hears the sound first inhibits the one that hears it later, giving to the brain the impression that the sound reaching the nearer ear is much louder than that arriving at the further. In theory, the time difference from a sound coming from ahead will be the same as that from a sound behind. In practice, sound shadow will separate the two. Also, as soon as the head turns, say to the right, an increase in sound in the right ear will indicate the source is behind.

When the ears are level, a bird will measure direction in the horizontal plane. To find vertical alignment, it will be necessary to tilt the head sideways. Yet small birds with only half an inch between their ear-holes, giving them only $\frac{1}{25,000}$th of a second to play with, cock their heads to listen to a sound so they must be using time-differences. On the other hand, an owl with its broad head may have ear-holes three or four inches apart, giving it a few thousandths of a second with which to find direction.

An owl has another advantage. It has internal ear-flaps leading sound into its ear-holes and the right-hand flap is one quarter of the ear separation higher than the left so that there is in effect a separation of the hearing in the vertical plane.[33] Thus the owl can find vertical direction by nodding up and down and can sense the horizontal alignment of a sound by turning the head from side to side. By this means, in complete darkness, it will strike accurately at a wad of cotton wool dragged over leaves to imitate the rustling of a mouse, even though the target has neither smell nor body heat.

It will be much more difficult for the ear that hears the sound first to inhibit the hearing of the other ear if the wave coming in only builds up very slowly, in other words, if it has a low frequency. Hence direction-finding by time-differencing requires high-frequency sounds. This is why the owl can catch anything that rustles, for rustling is a mixture of high-pitched frequencies. On the other hand, it will not be able to capture a live mouse at night if the little creature is moving over a hard floor and makes no rustling. Indeed, it cannot use its ears to hunt if the frequency of sound falls below 5,000 cycles a second.

Infra-red films have been taken of an owl trying to catch a kangaroo-rat, a remarkable little animal which lives in the desert without drinking water. The little rodent only comes out at night when the sand is cool and, for self-preservation, it has hearing organs bigger than its brains, with large cavities behind the eardrums to make sure the sounds are not attenuated by inside pressures. Its ears are extremely sensitive to frequencies above 1,000 cycles per second.

Owls have only survived as night hunters by being able to fly extremely quietly, with soft down on the edges of their wings to deaden turbulence, and also they fly slowly to reduce noise still further. However, they cannot avoid a very faint rustle at 1,200 cycles per second. The kangaroo-rat detects this and films have shown its ability to jump out of the way at the very last moment. It dodges snakes in the same fashion, for these reptiles rustle faintly at 2,000 cycles a second as they strike.

The kangaroo-rat, like all mammals, except whales and dolphin, has the advantage of external ear-flaps. These concentrate the sound waves and, by being turned, identify the direction from which the waves have come. A radar antenna has a parabolic shape which reflects incoming radio waves into a focus. A mammal's ear is similar except that one half of the parabola is generally taken up by the animal's head as shown in Fig. 26(a) and the surface may be shaped to alter the reception pattern from certain angles. In both radar antenna and ear, there is usually a secondary reflector close to the radio source or the ear-hole. This is very noticeable in human beings.

If we draw patterns to discover how the sound waves will be concentrated by a reflector, we naturally find that, the larger the reflector, the greater its ability to concentrate and the more strongly directional it will be. Generally mammal ears are long vertically and,

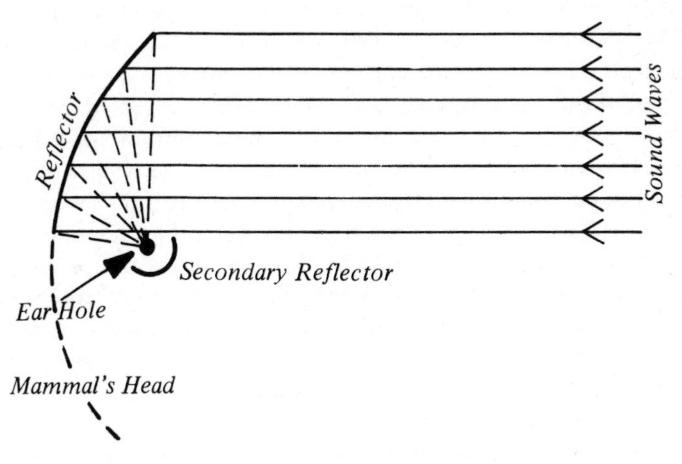

(a) Principle of Reflector

26. The mammal ear-flap

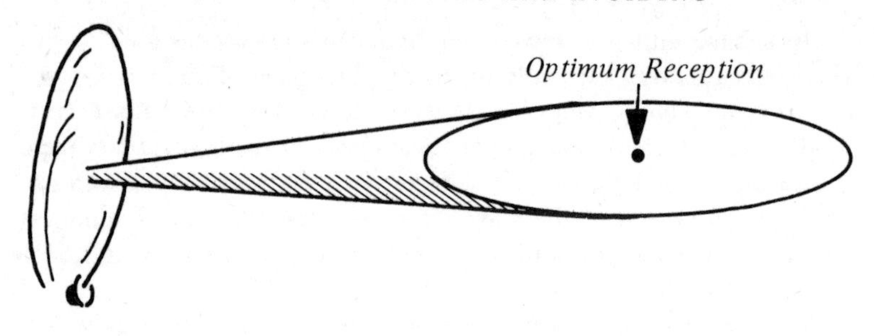

(b) Half Maximum Strength Pattern

as a result, the incoming waves are highly concentrated in the vertical plane. Fig. 26(b) shows a rabbit's ear and the consequent squeezing together of the sound waves to sense vertical direction more accurately.

In this particular figure, the mammal's ear is three times as high as it is wide. Therefore the sound waves are shown as being compressed three times as much vertically as horizontally, and three times as much sound will be picked up in the vertical plane. Naturally, the animal will turn its ears from side to side and tilt them forwards and backwards to find the direction of the incoming waves, but only one third of the 'bracketing' will be needed in tilt compared to turning. In addition, an animal such as a cat will 'prick' its ears and aim them together if it is interested in something it hears.

Size of ear is not the only factor. By halving the size of the ear and also halving the lengths of the waves, the pattern of concentration would remain in the same proportions and the accuracy of the direction-finding would be unaltered. Therefore, if the length of the waves be halved but the ears left as they were, the amount of sound they would concentrate would be doubled. Thus the precision of direction-finding depends not only on the size of the ear-flaps but also on the shortness of the sound waves. This follows the pattern of time-difference hearing which depends not only on the separation of the ears but also on the availability of high-pitched sounds, which are produced by short waves.

Most sounds are, like rustlings, a mixture of low tones and high overtones, the latter resulting from short waves. Therefore the accuracy of direction-finding will depend on how high are the frequencies or notes that the animal can hear. It is perhaps of interest to note that the drummings of rabbits, already mentioned as warning sounds, tend to be low notes and so the direction of the drummer is less easily determined by the predator. The warning calls uttered by birds are naturally higher in tone but they wax and wane which makes it difficult for an enemy to locate the source by turning the head from side to side to detect the direction.

The alarm calls used by small birds are intended not only to enable others to avoid the danger but also to encourage them to collect together and to drive the enemy away by 'mobbing', diving down, screeching and even pecking, or at least going through the motions. We have seen that some birds use special calls for birds of prey overhead and they may even produce sounds that indicate the type of predator. Jays for example will chitter if they see a squirrel on the prowl.

As a result of these calls to arms, starlings, blackbirds and the larger rooks, crows and ravens will attack hawks flying in the air. Sea-birds nesting in colonies will mob a fox or a human being, pecking at the head of the intruder. Other smaller and less aggressive birds may roost with these belligerents, gaining a measure of protection from them. Yet birds as small as tits will chivvy an owl sitting on a branch in a tree and will often drive it away. Tits even produce their own private warning sounds, hissing like snakes if disturbed in their holes.

Perhaps one of the most amusing examples of the use of sound for hunting is displayed by an African bird, the honey guide, and its bad-tempered ally the honey badger which, when hungry, will whistle softly to the bird. When the guide finds a bee's nest, it gives a special call and, if nothing happens, searches round for the badger and leads it to the nest by its call. On arrival, the badger breaks down the nest with its claws and exposes the honey for itself and the wax for its avian ally.

SONAR

Some animals produce their own sound waves and listen to the echoes to find out what is around them, a process analagous to radar and known as sonar. Just as the detail on a canvas depends on the fineness of the brush and the definition of a newspaper picture is according to the smallness of the dots, so the precision of anything painted by waves depends on their length. Therefore, in sonar, we look at wavelengths rather than frequencies.

From various figures which we have already quoted, we arrive at the following wavelengths:

(a) Light, 3 to 8 ten-thousandths of a millimetre,
(b) Bats in air, 3 millimetres, and
(c) Dolphin in water, 8 millimetres.

We now see that light has the potential to produce a picture ten thousand times more precise than sound. Yet we shall find a bat can catch a tiny insect on the wing in pitch darkness and a dolphin can detect a vitamin capsule dropped into the far end of a swimming pool. This is the magic of sonar!

Bats. Let us use the little brown bat to illustrate the problems involved in sending out sound waves and finding how long they take to return after bouncing off an object. When cruising around, the bat may send out little bursts of ultrasonic sound ten times a second. It may be noted that the squeaks children hear are not these impulses but are bats talking to each other or to themselves. Between the bursts, sound travelling at about 1,000 feet per second covers a distance of 100 feet so that echoes can be received from objects 50 feet away before the next block of waves goes out and confuses the issue.

If the bat picks up prey, say, 10 feet away, it needs to send out impulses as frequently as possible so as to follow any movements. As the distance has been reduced by a factor of five, the repetition rate can be increased by the same factor and the bursts of sound will come tumbling out fifty times a second. Thus the transmissions will rise from a low phutter to the high buzz of a bandsaw until eventually the impulses are pouring out at two hundred a second and the range is down to 2½ feet.

The bat is now running into problems. To pick up the echoes, bats have huge ears which may be nearly as long as their bodies, and they turn these ears to find the direction from which an echo is coming. To avoid upsetting the sensitivity of these receivers, a muscle pulls out the bone that links the eardrum to the hearing system for the brief moment while the sound is going out. At two hundred impulses a second, the muscle can no longer cope.

To show how important it is to disconnect the ears, Donald Griffin, who has done wonderful work with bats, suggests that a tape recording be made of a clicking inside a closed room.[34] The sound of the click will deaden the ears so that the echoes from the walls will not be picked up. However, if the tape is played backwards, the faint echoes arrive first and will be easily heard. Luckily, although the bat's ears cannot be disconnected two hundred times a second, the echoes coming back from $2\frac{1}{2}$ feet will be sufficiently strong for the bat to manage.

Even more fundamental is the fact that each little block of waves must last for a finite time. At these short distances, the echo from the start of each impulse will be back before the pulse has been finished and there will be no break in the returning sound. However, the bat produces each impulse not as a constant sound but starting at 100,000 cycles a second and dropping to a note only half as high, thus doubling the wavelength and giving a signature tune depicted notionally in Fig. 27(a). This makes it possible for the bat to distinguish an echo by the note coming back from a small part of the impulse. So the little brown bat homes in until the insect may be caught in its mouth. By this means, it will pick up two tiny fruit-flies in one second.

With such a staggering performance, one would imagine a comparatively large creature such as a night moth would have no hope. Not so! First of all, its body will have a soft outer layer, absorbing sound and making it more difficult to detect at long range. If it hears a bat cruising around, a night moth can tell the direction from which the sound is coming by the hearing organs just behind its wings. If the sound is blanketed when the wings are up, the bat is ahead and above or, if it happens when the wings are lower, the bat is ahead and below. In either instance, the moth will do a sharp U-turn whereas, if there is no blanketing, the bat must be aft and the moth hurries on. If the left hearing organ is getting most of the sound, the moth turns right and so on. Thus moths scatter when they hear a bat about.

The bat knows of this but continues on its way, wheeling drunkenly from side to side to make it more difficult for the moth to know where the danger is coming from. At twenty feet, the repetition rate of the impulses is building up and the moth knows it is in trouble. If it then simply drops straight down to the ground, the bat may have a fifty-fifty chance of capture so the moth goes into aerobatics, loops the loop, does stall turns and spirals.[35] Yet the bat has a final trick,

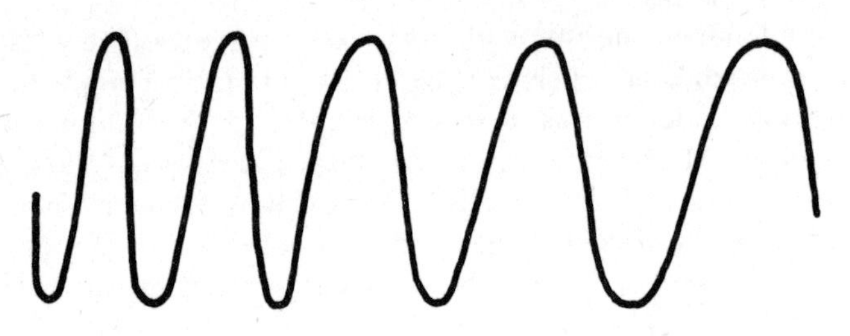

(a) Varying Notes of Constant Strength (Brown Bat)

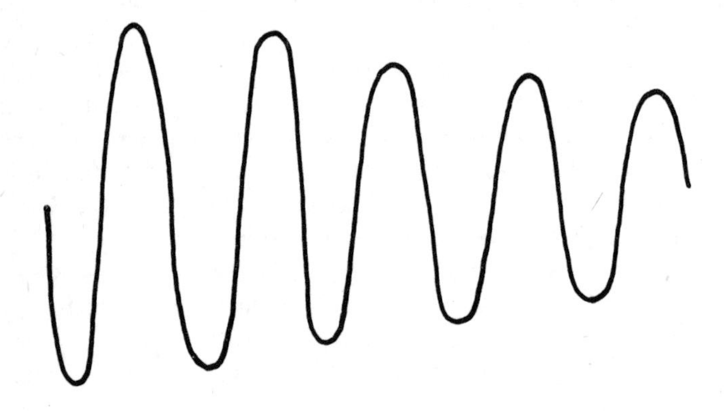

(b) Constant Notes of Varying Strength (Horseshoe Bat)

27. Bats' signature tunes

opening back legs at the last moment, hoping to catch the prey in the membrane between them or in his outstretched wings.

If the moth is unlucky, the little bat stuffs it into its mouth and, provided the moth is small, goes on transmitting through its teeth, for the bursts of sound emanate from the larynx. If the moth is large, the bat has to let the sound out through its nose and, as a result, the other moths are safe for the time being, although the sound is sufficient to avoid flying into trees or other bats.

The frog-eating bat which catches much larger prey has to use its nose to produce the sounds and has a reflector just above the nostrils to beam the transmissions at the target and also to keep them away from its ears at close range. The bat has a remarkable ability to differentiate between an edible frog and a poisonous toad of the same size and shape, apparently by means of small teeth made of skin and carrying taste or smell sensors, for it will snap at the toad but leave it quite unharmed. The horseshoe-bat also uses its nose and has large reflectors that confine the waves roughly within a cone of about twenty degrees each way and, incidentally, make it look fearfully ugly.

The horseshoe-bat transmits quite long patterns in which only the volume changes, the pitch remaining constant as shown in Fig. 27(b). The use of this extremely pure note has a special advantage. When a racing car goes by, the engine noise drops from a high scream to a low growl and, by measuring this 'doppler' change of frequency, the speed of the vehicle can be worked out. The horseshoe-bat is likewise able to tell by the change of note received whether an object is coming or going and roughly how fast. This makes it possible to pick up a moving object such as a dung-beetle against a static background. The change in strength enables the bat to use the signal even when the target is so close that pulses have to be sent out continuously. These bats also search by rotating one ear forwards and the other backwards, the greater horseshoe-bat working them at speeds up to sixty turns in a minute.

The fishing bats have downward pointing mouths and may well be able to produce bursts of sound strong enough to penetrate the surface of the water for a short distance and be reflected by the swim-bladders of fish. Whether or not this is possible, their sonars can certainly pick up the fins of very small fish breaking the surface when frightened by pelicans, with whom these bats often fly. Alternatively, they will detect minute fish 'larvae' a fraction of an inch long and a fraction of a centimetre across at ranges of a few feet.

In the chapter on visual hunting, it was mentioned that nasty tasting or dangerous animals advertise their presence by bright colours. Other creatures specifically warn bats by producing clicks at the same frequency as a bat's sonar. Bats are very fond of mealworms but will sheer away if a tape of their clickings is played at the same time. Inevitably, imitators have entered into the game. The tiger-moth, which is extremely edible, produces these warning clicks by contracting and relaxing muscles on the third pair of legs and these react on pliable plates and resonating boxes close to its body.[36] In addition, bats certainly recognise insects that sting by their wing beats, which also provides protection for the hoverfly.

Other animals. It is apparent that fish may be able to use their lateral lines to measure the time taken for the disturbance due to their motion to be reflected from objects such as the sea bottom. Perhaps the performance is comparable to that achieved by a blind man with his tapping stick and, indeed, many people with normal sight have 'felt' there is an obstruction ahead when walking in pitch darkness, probably by sensing echoes from their footfalls.

It is doubtful whether amphibians or reptiles use sonar but there is one bird, the penguin. As it swims, its feathers seem to produce bubbles which collapse and generate high-frequency clicks. Echoes coming from the swim-bladders of fish seem to tell the birds where the prey is but they can also find pieces of fish thrown by a keeper into a pool even in pitch darkness. They will completely clear the water and then come out but, when the lights are switched on, they dive back again to make sure they have missed nothing! Shrews, hedgehogs and many other small mammals use a crude form of sonar by making clicking sounds, and the hippopotamus can detect objects underwater when there is too much silt to see further than a foot or two.

Sealions and seals are able to pick up echoes from fish using sonar similar to that employed by penguins, the clickings being caused by the collapse of bubbles produced by their fur as they swim. It has been proved they can hunt better in dark turbid water for, while they can use their sonar, the fish have difficulty in seeing them. A sealion that has gone blind will remain in excellent condition for many years and so can have had no difficulty in feeding on its sole diet, fish.

However it is only certain fully aquatic mammals which have developed sonar comparable to the tiny bat's. These are the dolphin and the toothed whales which hunt fish and other marine creatures including those that live deep down in the ocean. A whale has even tangled with a submarine cable, presumably mistaking it for the arm of a very large octopus. These hunters produce powerful impulses of sound at frequencies far higher than a bat's and they have wide distances between their ear-holes which are lopsided, one higher than the other, like the flaps of an owl. So, when they are using sonar, they weave their heads from side to side and also nod up and down.

Baleen whales also use sonar but of a much lower frequency to detect, at long range, the great layers of krill on which they feed. Since these layers are static in the water, there is not the same requirement for asymmetrical ear-holes. It is naturally difficult to experiment with animals as big as whales and so most of the work has been undertaken with dolphin, one of which is pictured in Fig. 28. These delightful creatures seem to live quite happily in an oceanarium and probably their sonar is comparable to that of the toothed whale.

28. The delightful dolphin

In cruise, the dolphin sends out sounds at about ten times a second using its melon, the bulge above the mouth, and also the mouth itself. As with a bat, the rate at which these sounds are produced increases as the target is approached, going up to four hundred a second at which time interval the object would be a couple of yards away, due to the high speed of sound in water. In any event, dolphin catch fish with no difficulty even with pads over their eyes but will not allow the melon to be obstructed. Furthermore, their sonar can distinguish the texture of an object for they will home onto a piece of fish and ignore a gelatin capsule of the same size and shape. One dolphin named Kae could even signal when a cylinder was dropped into her pool.[37]

SUMMARY

1. *Touch.* Not a major navigational aid. Mole's main sense.
2. *Vibration.* Widespread. Fish and spider's main sense.

3. *Hearing*
   (a) *Insects.* Various types of ear on legs and bodies. Direction by shadow (wings etc.). Moths detect very high notes.
   (b) *Fish.* Swim-bladders, some linked to internal ears by bones. Hearing poor. Sounds used for mating.
   (c) *Amphibians and reptiles.* Eardrums open one side. Hearing moderate.
   (d) *Birds.* Ear-holes, direction by time differences. Good hearing over moderate range. Sound used for social purposes.
   (e) *Mammals.* Direction by large ear-flaps and time differences. Good hearing over wide range of notes.
4. *Sonar.* Employed by seals and penguins but developed by
   (a) *Bats* using short 'coded' transmissions of very high frequency,
   (b) *Toothed whales and dolphin* using very short transmissions, at frequencies even higher than bats', and asymmetrical ear-holes.
   (c) *Baleen whales* using low-frequency sonar.

5
Chemical and Electrical

SMELL AND TASTE

The olfactory sense influences many activities in the animal world. Above all, it is used to hunt for prey, to find a mate or to detect predators. Also, it identifies and binds together the members of a community that live in the same nest or hive or roam about together in the same pack or herd. Yet, in spite of its importance in the animal world, exactly how it works seems still to be in doubt. Even the division of scents into categories is a matter for dispute, the primary odours postulated by experts varying from eight to thirty-two though less than a dozen different types of sensor have been identified, each probably responding to several scents in different ways.

Detection of a chemical at a distance is obviously navigational but tasting is only possible when an object has been reached. However, just as touch, vibration and sound are difficult to separate, so are smell and taste. Even when they are distinguished, a very pungent odour will be detected by taste buds and, as gourmets well know, taste is greatly influenced by the smell of food. However, as a rule, taste involves responding to chemicals soluble in water whereas smell is affected by gases which are not dissolved in water though they may be, like most gases, in suspension.

There are two particular chemicals that are generally detected by animals. The first is water in the form of humidity in the air. Insects and mammals may well be able to smell water as such, while certain other creatures may carry specialised sensors to detect whether there is an excess or a deficit of moisture around. The second chemical is carbon dioxide, which is detected by a wide range of creatures. Human beings, for example, can identify a 'stuffy' atmosphere and they carry inside their brains sensitive carbon dioxide blood samplers, which stimulate an increase in the breathing rate when the proportion rises above a certain level.

Invertebrates. Single-celled creatures are sufficiently sensitive to chemicals to select some foods in preference to others. They may be able to make the choice at a distance, perhaps detecting the concentration of carbon dioxide by using very fine hair-like sensors. By similar attachments to mouth and skirts, the jelly-fish smells out its prey. Flatworms, with detectors on their heads, crawl upstream when attracted by a scent.

A starfish, with sensors on its extremities having a concentration of 4,000 cells per square millimetre, will twitch as it passes over a spot on the sea-bed covered by the tide and then dig down four inches in shifting sand to a clam it has smelt.[38] Meanwhile, the clam, with sensors which react to the odour of a starfish or a predatory mollusc when its shell is open, will already have closed up. Squids have detectors on various parts of their bodies and crabs smell with main and with secondary antennae and sometimes with other parts such as gills and mouth. A freshwater snail, using its foot to smell a leech, will drop to the river bottom for safety.

Leeches and earthworms appear to carry sensitive areas in the fronts of their mouths to detect food and the latter, when injured, is said to produce a warning chemical which frightens other worms away for long periods of time, apparently lasting sometimes for weeks. Snails find their mates by smelling their slimy tracks and, if hunters, they pursue other snails by the same means. Millipedes and centipedes also rely partly on scent to find food and mates.

Many spiders recognise the scent in the line leading from a web to its owner by means of receptors in their feet and identify the occupant as belonging to a certain species and being of a particular sex. They can avoid going onto the web if the owner is of a different type or the same sex. Even if the scent is right, a small male may be sufficiently attractive to a large female to be tasted as well as smelt and subsequently devoured. A similar fate often awaits the male flower-mantis.

A male mosquito confirms that a female is of the right species by means of smell. Blowflies have hollow hairs on their feet as well as in their tubular 'mouths', with half a dozen sensors in each reacting to water, certain sugars and a few particular salts. In these lower animals, the sensors for taste and smell seem to be identical, the effect of a chemical liberated at a distance being merely less than the same chemical in contact. There are also certain creatures that do not seem to be able to smell at all. Crickets and 'silver-fish' are said to be descended from ancestors which lived 250 million years ago, before flowering plants had appeared and when there was no requirement for the smelling powers of the more modern butterflies and bees.

Insects smell mainly with antennae and parts of the mouth, using minuscule hairs, pegs, pits or plates which are also found on other parts of some creatures. The honey bee carries a million cells in its antennae which give it a sensitivity to the scent of flowers at least as good as a human being's and, in addition, makes it extremely responsive to the smell of its queen, which does not register on human noses at all. Also, when a bee or an ant returns after a successful forage, the others cluster round it to smell the pollen or the food which has been brought back.

A fire-ant, finding food too heavy to bring back to the nest singlehanded, lays a trail of scent with its sting using a directional sequence, increasing the pressure and the amount of scent and then lifting off for a moment before starting again gently. Other ants find the scent and recognise its direction and they will reinforce it with their own trails when they return, provided some of the food is left. The scent evaporates quickly and so ants are not confused by a multiplicity of trails leading from various sources of food which have been exhausted.

Ants are peculiar creatures, living in a world of taste and smell and milking aphis for their honey-dew. They also recognise the caterpillar of the large blue butterfly by taste, for it produces a liquid they like so much they will carry the caterpillar to their nest and feed it on their own larvae. When it becomes a butterfly, they let it go.

Insects also use chemicals for defence and some carry poisonous stings. A bee may mark the spot stung with a special scent so that others can repeat the performance. Ground beetles may even discourage predators by shooting acid at them. The bombardier beetle ejects a corrosive liquid, heated almost to boiling point, which evaporates into a little cloud and makes a small popping sound. Social insects use a particular scent to identify themselves as members of a nest or hive and the absence of this provokes immediate attack. So an ant which has lost its antennae will attack the others in its nest, taking them for intruders. Such insects produce alarm chemicals when they or their homes are threatened, the scent acting as a battle cry.

Creatures that pollinate flowers have had to learn to differentiate between various scents but generally insects respond in specialised ways to a limited number of chemicals. Many take action on detecting carbon dioxide. Bees, for example, start to fan with their wings when they are in their hives. Mosquitoes use antennae to head straight for the source, knowing it may emanate from a human body. Sucking beetles do the same underground, for vine roots give off carbon dioxide.

Insects that lay eggs in others rely largely on smell. A chalcis beetle, confronted by a heap of a hundred thousand grains of which a hundred and twenty had been occupied by grain weevils, laid its eggs in all but four of the grains even though some were buried over a foot deep in the pile.[39] These creatures will lay an egg in a grain which already has a hole in it provided they do not smell that another female has walked over it. The ichneumon fly, able to hear larvae moving about in wood, employs scent to identify the few suitable to act as hosts and which are not already occupied. When subsequently a female fly hatches out, several males of slightly different species will be waiting for her but, by using smell, only the right one will mate.[40]

These are the workings of highly specialised senses of smell. Yet the most specialised of all, the most sensitive and yet the most restrictive, appears in certain male moths. The female silkworm moth has generalised smelling sensors like other insects. When ready to breed, her body produces about a hundredth of a pin's head of chemical of a very specific type and she releases this into the air.

The male, a specialist so dedicated he can neither eat nor drink,[41] carries two tiny frond-like antennae illustrated in Fig. 29, each with 20,000 sense cells and 35,000 nerve fibres mostly sensitive only to the

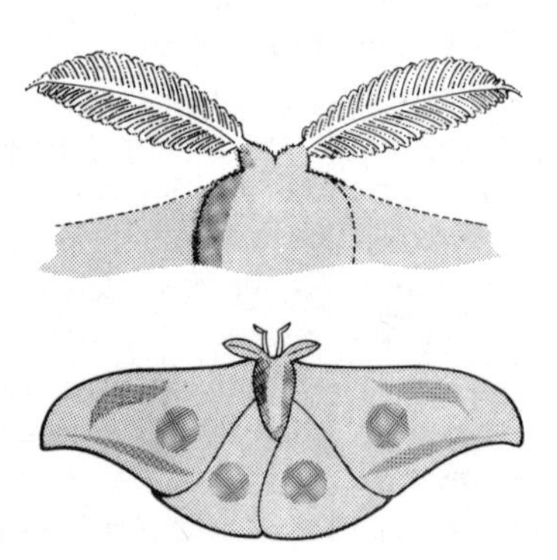

29. Frond antennae

female chemical and to no other, detecting the molecules perhaps by their shape or by their molecular vibrations. Thus he will sense this incredibly minute amount of perfume diffused through a volume of air of many cubic miles. He homes in, first simply travelling upwind, and as he gets closer compares the scent gradient at each antenna, simultaneously giving off an odour to deter any rivals. So his brief life ends.

Vertebrates. In vertebrates, the organs for the detection of flavour are differentiated from those that react to scent. The former rely on taste buds which may be replaced at intervals, every ten days in certain mammals, and may produce reflex reactions, such as salivation. They are to be found on the upper surface of the tongue and, notably in fish and amphibians, on lips, cheeks and throats while many fish have them in their outer skins, particularly on their gill covers, stomachs, fins and tails. The olfactory sensors on the other hand tend to be concentrated in at least one wall of the nasal cavity, water or air being moved across them, and the nerves from them run to a special lobe of the brain, thus involving analysis and memory intimately with the sense of smell.

In addition, the sense of smell is supported by an area in the roof of the mouth, found in amphibians, lizards and particularly in snakes, which have poor nasal sensors, and it also appears in some mammals. Instead of relying on air or water flow, the moist tongue picks up molecules outside and carries them to the sensitive area in the roof of the mouth, a snake having two pits into which the two prongs of its forked tongue can be inserted.

Fish have an extremely keen sense of smell supported by large lobes of the brain which, in the shark, are enormous. Even a little minnow can be taught to distinguish the scent of over a dozen different types of fish. Fishes that lay eggs in the nests of others operate by smell; mouth-breeders, which carry their young about with them, can tell their own offspring by their odour. The noses of fish consist of tubes, generally paired, leading from just in front of the eyes and typically carrying as many as a hundred thousand cells per square millimetre of surface, these sensors being frequently supported by those on other parts of their bodies.

Fish work upstream when they detect a scent that means food or a mate and they can sense minute variations of the chemicals in the water round them. An eel has small eyes but, with nearly a million sensitive scent detectors, it must have a capability comparable to a tracker dog. Moray eels hunt by smell, but the intelligent octopus lets off a cloud of sepia, not to prevent the eel from seeing but to upset its sense of smell. Sharks are excited by the smell of blood which they detect at extraordinary distances in concentrations much less than one in a million.

A story is told of a large and hungry minnow that came upon a shoal of smaller minnows and was tempted to cannibalism. It caught one of its smaller relations but almost at once it slunk away as if conscience stricken, although the rest of the minnows did not react at all. When the large minnow ruptured the skin of its victim, it caused the release of a so-called shock substance, which frightened it but not the other minnows, who were too young to recognise the odour.

Nearly all fish, particularly the fresh water varieties, emit shock substances when their skin is broken and this causes panic in their own species. Among the exceptions are eels, grayling, pike, stickleback, trout and, at sea, herring, mackerel and sardines. Toads, tadpoles and water snails also give out shock substances. In certain instances, these substances alarm other sorts of fish, for example a wounded carp will frighten a salmon. Unhappily, predators, particularly sharks, have learnt to swim towards the scent of a shock substance, converting a selfless emission into a threat to others.

Reptiles, particularly crocodiles and alligators, have well-developed nasal cavities except for turtles which nevertheless have a good sense of smell, the snapping-turtle being the bloodhound of the water. Snakes identify prey by scent, using the forked tongue and the pits in the roof of the mouth, and some cannot track their victims if this organ is injured. King snakes follow the scent of rattlesnakes which are warned by smelling their predators. In a similar way, male snakes follow females but may not court if their nostrils are taped over. Other snakes give off offensive smells to discourage their enemies.

Birds rely mainly on sight and sound but many, including duck, quail and sea-birds, particularly albatross, have large olfactory brain areas and must therefore have a good sense of smell. The kiwi lives on food found underground and has a long beak to dig it up, using its acute sense of smell and nostrils in the end of its beak! In general, however, birds have rather simple nasal cavities with less numerous sense cells than fish or mammals.

Mammals, short-sighted but sharp of hearing, use their noses mainly to find food. That famous French delicacy, the truffle, buried a foot underground, can be detected by badgers, bears, deer, field mice, rabbits, squirrels, wild boar and wild cats. Even more sensitive to its odour are dogs, goats and, above all, pigs who will smell it fifty yards downwind. No wonder hunting mammals bury their kill and rely on their noses rather than memories to find them again. Fortunately, baby rabbits, buried by mother in the sand when she goes out for food, have so little scent they escape unless a fox passes directly overhead.

Apart from the aquatic whales and dolphin, which either have very small olfactory brain areas or none at all,[42] and to some extent the seal and walrus who feed in the sea, mammals have complex and extensive nasal cavities, often with small organs in the roofs of their mouths similar to those used by snakes, though these are absent in bats, monkeys and human beings. Scent dominates feeding and it has social significance, being used to mark lairs and territories, while members of a flock of sheep identify one another by their smell. Also, from the lowest forms of mammal to the highest, scent is used by males and females to attract each other.

However man, the so-called 'highest' mammal, uses his or her nose mainly to warm the air going into the lungs and makes only a limited use of the sense of smell though, with practice, human beings, especially those with dark skins, can perform remarkably well. With 40 thousand cells per square centimetre and, to quote Dr. Maurice Burton, an area only the size of a postage stamp, man naturally compares unfavourably with the dog which has twice the density on fifty times the area, producing an incomparably greater sensitivity. Furthermore, the olfactory bulb in a dog's brain, which analyses the scents, is far larger than a human's. Rabbits, with cell densities even greater than dogs, hedgehogs and rhinoceros are similarly endowed, and that delightful anachronism the elephant is said to have a sense of smell as good as a moth's.

Mammals, particularly the vegetarians, give off an odour when adrenalin is pumped into their systems. This enables others to detect when they are frightened. It is well known that animals can tell when people are afraid but whether it is an improvement on the systems of fishes, which require an actual injury to produce a shock substance, is hard to tell. On the other hand, predators take great care to remove all scents that could give them away, which is why cats clean themselves scrupulously and bury anything that stinks. At the other extreme, some mammals defend themselves by horrible smells. The skunk is able to cause choking and even temporary blindness at a distance of ten feet.

Finally, let us consider what a tracker dog can do, not that its sense of smell is unique in the animal world but rather that its performance has been so well documented. The sense is certainly remarkably acute. A dog has been known to track a donkey which had travelled over rocky ground four days earlier. It will pick up the minute amount of butyric acid that has percolated from a human foot through a rubber boot. It can find coffee, tobacco, opium or explosive concealed in luggage or hidden in crates and, in parts of Europe, it is used to detect gas leaks underground.

Not only is the nose of the dog sensitive, it is also extraordinarily selective. It has been claimed that a highly trained human perfume expert can probably identify over a thousand different scents but, to a tracker dog, there are as many odours as there are dogs, cats, rabbits and people and all are different. A dog has followed the scent of one man in a party of a dozen walking in single file and each stepping into the footprints of the man in front, and has continued to follow even when the party split into two. A dog will even distinguish one identical twin from another though their scents are so much alike that, if one should pass by on his own, he will be followed by mistake for the other.

When a dog picks up a track, he will follow it for about twenty yards and then either continue or back-track in the opposite direction. What he has done is to recognise those elements of the scent that evaporate the quickest. If, on his exploratory twenty yards, the quickly evaporating scents have dispersed to a greater extent, he knows they must have been made earlier and so he reverses his path. If they have dispersed less, he knows he is on the right line. Yet the difference in time involved if he is tracking a human being travelling at walking speed is that, in one instance the scent was laid down a quarter of a minute before and in the other a quarter of a minute afterwards. This is perhaps the greatest marvel of all in the sense of smell.

Fig. 30 summarises, in a greatly simplified form, the methods used by animals to smell and to taste. In addition, the sensitivity of skins of aquatic animals to certain chemicals is retained in a vestigial form by

<table>
<tr><th>Animal</th><th>Smell</th><th>Taste</th></tr>
<tr><td>Mammal</td><td rowspan="3">Nerve Cells in Nasal Cavity
Air drawn over cells into lungs.

Lizard & snake also use tongue to carry chemical to roof of mouth.</td><td rowspan="3">Special Cells
mostly on upper surface of tongue</td></tr>
<tr><td>Bird</td></tr>
<tr><td>Reptile</td></tr>
<tr><td>Fish</td><td>Paired Nostrils
Water flows over cells as fish moves.</td><td>Cells
in or near mouth & on fins, tail, etc.</td></tr>
<tr><td>Insect</td><td>Generally Antennae</td><td>Mouth Parts, Feet and Outer Skin etc.</td></tr>
<tr><td>Simple Creature</td><td colspan="2">General Skin Sensitivity</td></tr>
</table>

30. Smell and taste sensors (greatly simplified)

higher animals at the junctions of their inner linings with their outer skins, for example, the mucous membranes of eyelids and lips.

ELECTRICAL

When discussing the detection of magnetism, it was noted that muscles are activated by small electrical currents. Electricity is therefore an element common to all animals. However, air is an insulator, whereas electricity can pass through water provided it is not absolutely pure. Hence the only creatures that appear to employ it are fish, amongst which there may well be three hundred species that have some capability. The sensors consist of millimetric pores or small pits with a thin layer of conducting jelly at the bottom leading to an electrically sensitive cell below.

Detectors such as these are mainly carried on the heads of fish, including sharks, but sometimes also on their backs and their bellies. In certain fish they can pick up one thirtieth of a millionth of an amp of current per square centimetre, a quantity so negligible as to be virtually undetectable by engineering standards. By using these sensors, marine predators, particularly those that live at great depths, can detect the electrical currents in the nerves of fish close by and these are produced even when the prey is keeping quite still. It is also believed these pits can pick up the electrical currents in the Earth below which presage an earthquake leading to a 'tidal' wave.

Tunny may well have similar sensors, for fishermen catch them by passing a direct current of quite a small voltage between two electrodes immersed in the water. Skate and certain other fish may be able to produce currents of this nature in addition to the sensors carried in their heads. In their tails, the muscles have been converted into a large number of small, flat round plates placed on top of each other. We saw on page 84 that the breaking down of ATP is used not only to drive muscles but also to generate light. By breaking down this same chemical, these muscle developments can produce an electrical voltage. Thus the tails of skate can produce an electrical current which may well be used to attract certain fish.

Rays are large flatfish which, unlike soles and plaice that turn sideways, have developed by, so to speak, being flattened from above. They have powerful pectoral fins each side with organs consisting of several hundreds of these flat, round muscle plates stacked one above the other in a number of vertical columns. These produce a potential difference positive on top and negative on the bottom of about fifty volts, enough to stun a small fish and, indeed, the ancient Romans used the ray as a cure for gout. The torpedo-ray can generate 200 volts and though, like all rays, it is a sluggish creature and an awkward swimmer, fish such as salmon have been found in its stomach which it could not possibly have caught by conventional pursuit.

Naturally the ray, like all fish that generate their own electricity, has an insulated skin and nerve channels well protected from the voltages. The ray may also use its organs to find out how well the water carries current to the sensors in the head. Salt water, with its additional impurities, is a much better conductor of electricity than fresh water and the ray could detect this and know whether it should swim out to sea or go back closer inshore to find the water to its taste. Perhaps we should have included this under chemical detection.

The electric eel, which is not a true eel but a knife fish, is a much more powerful creature and has three separate organs. The main one is ten feet long and fills up most of its tail, taking up nearly half its body weight. It consists of no less than seventy columns each consisting of 6,000 to 10,000 plates but laid horizontally with their front ends positive and rear ends negative. These produce a series of shocks between 350 and 600 volts which is two and a half times that of the highest domestic voltage, each shock lasting a five-hundredth of a second and repeated two hundred times a second. At a considerable distance this will stun a fish which, unless it is also electrical, will not have an insulated skin for protection but a body that is a good conductor of electricity. The shock will even be enough to knock over a man or a horse that gets too close.

At the tail end of this main organ is a smaller one which the 'eel' probably uses to detect what is around it, because it sends out electrical impulses only when the fish moves, and so must have some navigational purpose. A third organ below may well be used to attract fish.[43] The electric catfish, a plump three-footer with a large round tail, produces 450-volt shocks for self-defence or to stun prey. Star-gazers, tropical fish with heavy heads, have organs below their eyes which have developed from eye muscles and produce about fifty volts for hunting or to discourage predators.

The most remarkable electrician of them all is perhaps the gymnarchus, first studied closely by Dr. H. Lissmann of Cambridge, England.[44] The fish inhabits the muddy bottoms of turbulent and

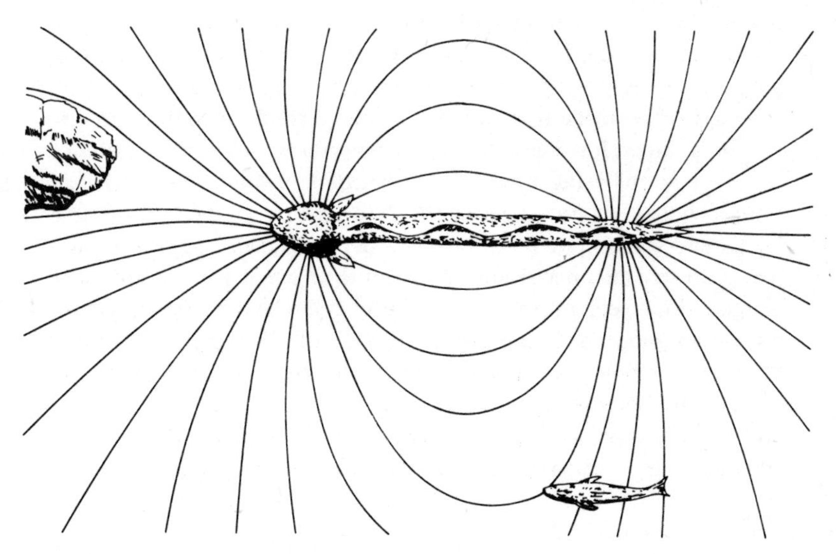

31. The electrical gymnarchus

torrential African rivers so that it can see nothing. From eight thin spindles, each consisting of 150 to 200 plates, it produces pulses of between three and ten volts about three hundred times a second which are positive at the front end and negative at its tail, thus building up an electrical field as suggested in Fig. 31. Mainly on its head, but also on its body, are extremely sensitive detectors.

Although relatively stupid, the gymnarchus has a brain two or three times as big as that of a normal fish of its bulk, the huge computer being almost entirely devoted to analysing the signals from its electrical sensors. The presence of prey in the form of a fish, a good conductor of electricity, is marked by a bunching of the lines of electrical force whereas an obstacle such as a rock, which is an insulator, produces a spreading of the lines, typical effects being shown in the Figure. If another gymnarchus should approach, it will produce a field of its own and, when this is detected, the two fish change their pulse rates and engage in a battle of electricity until one is discouraged and departs, to leave the field to the other.

The eyes of the gymnarchus are useless in the conditions in which it lives and probably do little more than recognise the change from day to night. The body must not bend, otherwise the electrical pattern will be distorted. So the fish swims as stiff as a ramrod, moving either backwards or forwards by undulating the fin running along the top of its body and indeed it often explores tail first. It is because of this ramrod necessity that we know the organs in the tail of a skate, which is waggled when the fish swims, cannot be used for finding prey.

The gymnarchus has relatives that perform in a similar fashion, using huge analysing brains and swimming stiffly. The American knife-fish has its fin underneath. It produces a pulse of variable frequency which may be as low as 2 per second in sluggish waters, going up to 1,600 per second in torrents or when the fish gets very excited. The elephant fish has similar characteristics, operating with two organs close to the tail and, as usual, with the front end positive. However, although as a rule the gymnarchus and its relatives are solitary and not intelligent, certain elephant fish appear to be sociable and to communicate with each other electrically and, what is more striking, some elephant fish, like the dolphin, are playful!

SUMMARY

1. *Smell.* A major sense used by all creatures including:
   (a) *Insects.* Specialised and ultra selective for mating. Generalised for pollen gathering.
   (b) *Fish.* Extremely keen sense. Shock substances produced.
   (c) *Reptiles.* Used to identify prey.
   (d) *Birds.* Relatively simple nasal cavities.
   (e) *Mammals.* Main sense used for hunting.
2. *Electricity.* Used only by certain fish to:
   (a) *Detect* either
      (i) *Passively* by picking up currents in muscles of prey, or
      (ii) *Actively* by generating a field, and sometimes to
   (b) *Stun* by shock.

PART THREE

Finding the Way

We have discussed how animals may use sight, hearing, smell and other senses in order to hunt, to find a mate and to avoid being caught and eaten. Animals are seen to develop these senses in specialised and often remarkably narrow ways to meet their needs. A bat catches a moth by miraculous sonar developed to a most sophisticated extent. A male silkworm moth finds his mate by detecting molecules of her scent, micro-sampling far beyond the capabilities of man. A gymnarchus fishes in troubled waters by producing an electrical field and sensing its deflections by detecting minute changes of voltage immeasurable even by modern standards.

Often two or more senses are used. The worker bee, using the Sun for guidance, flies along a route given by another bee until it smells the pollen its mentor brought back. The wild dog tracks quarry by scent until it can chase visually. Yet even these senses appear to be employed in a definite sequence according to when each is needed. For it appears that hunting and avoiding, exciting and terrifying though they may be, are fundamentally stereotyped activities.

In complete contrast, the methods used by animals to find ways are as diverse as their routes. They use whatever information there is to hand. They depend for their migrations on unconsidered trifles which have made it possible for their ancestors, by a long and arduous process of natural selection, to develop their ways around the world. The type of question we have to ask is not 'how does a bird navigate?' but 'how did the little wild Manx shearwater, taken from its nest on the Welsh coast and carried 3,000 miles across the sea to Boston, Massachusetts, manage to fly straight back home so as to arrive only twelve days later?'

So-called primitive peoples also seem to have relied on specialised methods. Spencer Chapman, the explorer, was with a party of Eskimos, paddling back to their home fjord along the Greenland coast in their frail kayaks, when dense fog clamped down. To his amazement, the paddlers were unconcerned, laughing and splashing, but keeping a safe distance from the high cliffs by timing the echoes. An hour or so later, they suddenly all turned at right angles and almost at once the home fjord loomed up ahead. On every headland, a cock snow-bunting was staking his claim, each with a slightly different song. When the paddlers heard the call of the bird on the headland leading to their fjord, they turned inwards.

Nobody would suggest that the student navigator of today should study bird calls. Ye this story emphasises that modern marine, air and space navigation has gone away from the particular to the general. Mariners, airmen and astronauts use standard universal systems that enable them to find their ways to anywhere they want to reach. As a result, they may concentrate on radio and inertial navigation but still employ astro-navigation and visual information and this is illustrated in Fig. 32. It is true that in the past seamen have 'smelt' land. It will also be noted that, although human navigators use the Earth's magnetism and ocean currents or winds for working out a future

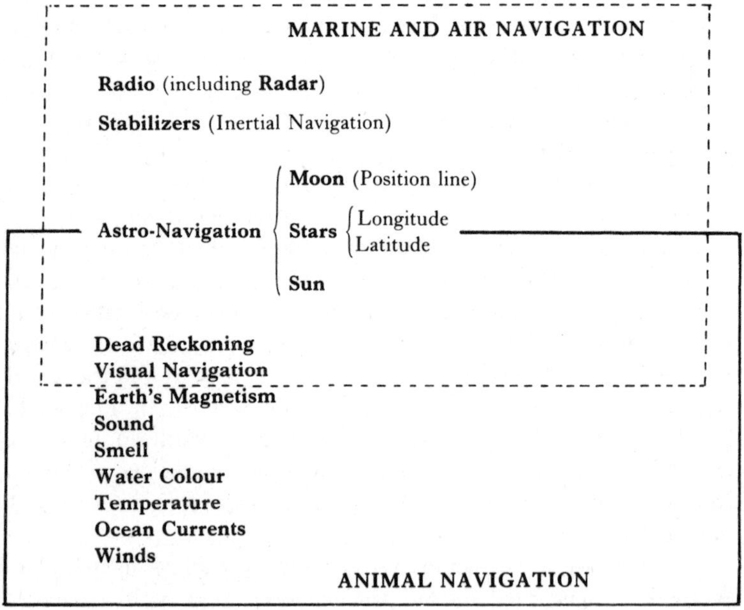

MARINE AND AIR NAVIGATION

Radio (including **Radar**)

Stabilizers (Inertial Navigation)

Astro-Navigation
- **Moon** (Position line)
- **Stars** (Longitude, Latitude)
- **Sun**

Dead Reckoning
Visual Navigation
Earth's Magnetism
Sound
Smell
Water Colour
Temperature
Ocean Currents
Winds

ANIMAL NAVIGATION

32. Finding the way by humans and animals

12. *The Arctic tern*, the size of a thrush, with a rather fluttering flight, travels 25,000 miles between Arctic and Antarctic each year. The tern follows remembered coastlines. *Stephen Dalton, Natural History Photographic Agency*

13. *The blackcap* has a good bump of direction. Imprinted on its tiny brain is probably the general rotational pattern of the night sky. *S. C. Brown, Aquila Photographics*

14. *Eels at the elver stage*. Eels can orient themselves probably by the Sun, and, if displaced fifty miles or so from home waters, are generally able to return to them. *Heather Angel*

15. *Locusts*. Migrating locusts may number ten thousand million in a cubic mile of air; they may travel several thousand feet above the ground and cover many thousand miles in a month. *Anthony Bannister, Natural History Photographic Agency*

situation, position is not found directly from these separate aids. Also, sound signals are used at sea but to avoid accidents rather than to find the way.

On the other hand, Fig. 32 emphasises the wide range of aids which animals employ. Even visual information is used in a very broad sense and not only for identifying 'pin-points' but also, and perhaps even more frequently, for the steering guidance afforded by topographical features such as coastlines, rivers and mountain ranges. We shall therefore start by explaining why animals do not employ the aids so greatly favoured by modern navigators and then discuss how they may use other sources of information. We can afterwards turn to the journeys they make and leave it partly to the imagination to suggest which aids they are likely to favour or whether they are relying on methods of which as yet we know nothing.

6

Systems

UNLIKELY AIDS

Radio. When concentrated by radar dishes, radio waves have been known to frighten birds. Flocks of geese, gulls, scaups, scoters and certain songbirds have scattered when picked up by a beam and have only formed up again when they have passed it by. Indeed, some birds have flown in circles and have behaved most peculiarly. Yet others, including crows, have not reacted at all. However, as most animals seem to detect magnetism, we might expect them to sense powerful electro-magnetic waves. It is even said that some human beings possess this ability.

So far as we can tell, no natural objects on the Earth except flashes of lightning transmit radio waves and attenuation by the atmosphere seems certain to reduce floodlighting from sources in space well below the strength at which they could be employed, even by very large and cumbersome antennae. Nor is it likely that animals could transmit without being picked up by modern radio receivers. Of course, passive reception enables a ship to find its position by direction finding. But animals are unlikely to carry mental pictures of the positions of radio stations and still less to be able to identify frequencies and call signs.

Nevertheless, we have to be careful. As we know, some birds at least can sense radio waves and therefore they may be assisted in recognising the locality of a transmitter particularly at night and when the transmitter produces a rotating beam, like a coastal lighthouse. So we have to accept that birds may well use radio stations as landmarks and it is reasonable to expect that other animals might use them in a similar fashion. But surely not small insects; for radio waves transmitted by ground stations range from inches in length to distances of miles from crest to crest.

Coriolis. It was H. L. Yeagley who first suggested Coriolis might be used by creatures to find the way.[45] This acceleration is caused by the Earth's rotation and is a maximum at the poles but nil at the equator, where it becomes a tilt in the vertical which involves memory and so is not to be registered by animal stabilizers. However, because the acceleration depends on latitude, it seemed animals might be able to sense how far they were from the equator, or rather from the nearest pole.

In 200-mile-an-hour aircraft of the nineteen-forties, Coriolis acceleration in temperature latitudes was found to deflect the vertical, measured according to the spirit level in a bubble sextant, by around $\frac{1}{15}°$. To introduce a deflection comparable to that of the 1° vertical of an animal would therefore require a speed of 3,000 miles an hour! Even if its vertical were so sensitive that the creature could move more slowly, the acceleration would be swamped by irregularities in the direction and speed of travel, unless an extremely sophisticated auto-pilot were fitted!

Inertial navigation. We have seen that animals and people are able to find their ways in the dark over very short distances by summarising the information from self-sensors in the limbs that move them about. It had been thought that inertial navigational systems could similarly integrate accelerations into speeds and convert speeds into distances so as to enable the passage from a known starting point to any other place to be recorded.[46] Unfortunately, it was found this would only operate over an extremely short period of time, owing to inevitable inaccuracies in measurements of accelerations, in the two processes of integration and in the times associated with each integration.

As a result, inertial systems today measure position by recording the change in the direction of the vertical when travelling from one point to another on or above the Earth's surface, which is why they cannot be used to find position in space. To achieve the required accuracy of a few miles on the surface of the Earth, an inertial navigation system fitted to an aircraft uses gyroscopes which remember the vertical at the starting point to within $\frac{1}{60}°$ for each hour of travel, and even more precise instruments are used for ships and submarines.

Animals cannot record the vertical to much better than 1° and, though this would only increase the consequent error by a factor of sixty, which might be acceptable, memories of signals passed by labyrinths to the brains even of higher animals only last for seconds rather than for hours. There is perhaps one even more convincing reason why animals cannot use inertial navigation. To reach a chosen destination, the system has to know the direction of the vertical at this final point. No creature could find out by using its stabilizers the direction of the vertical at the place it wants to reach.

USING SUN, MOON OR STARS

The Sun. But the little wild Manx shearwater *did* have a way to remember the vertical at its nest! Fig. 33(a) is a copy of Fig. 13(b) in which the hemisphere represents the sky. On it, the thick line indicates the motion of the Sun, which the bird is able to follow, and the dotted line shows this motion extrapolated to find the position of the local 'midday Sun'. The thick double line is a reminder that the 'midday Sun' is the closest point which the Sun reaches compared to the nest.

On page 48, we saw why a bird might prefer to estimate the time at which the Sun would reach this 'midday' position. So the little Manx shearwater transported to Boston, Massachusetts, with its internal clock still set to Welsh time, could look at the Sun and, allowing for the motion and the time, find the 'midday' point in the sky which the Sun would reach. This would, of course, be the point at midday in Wales.

From this 'Welsh midday' point, the vertical at the nest would be obvious. It would be at right angles to the Sun's path and at the same distance from the 'Welsh midday' Sun as when the bird was on its nest. Of course, the bird will be, so to speak, looking out at the hemisphere of the sky from the centre and there will be no distortion of the right angle such as appears from looking at it sideways in Fig. 33(b). Once the vertical at the nest can be sensed, all the bird has to do is to fly towards it. The bird would then follow the shortest route home, known to navigators as the 'great circle', provided it were not unduly deflected by cross-winds when flying over the sea.

When a fielder goes to catch a ball hit straight up into the air, he instinctively imagines a vertical through the ball and runs towards it. Yet he does not shut his eyes and hold out hopeful hands but makes a

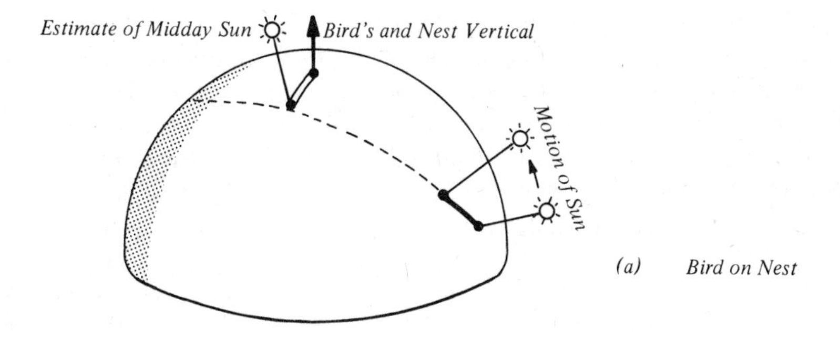

(a) Bird on Nest

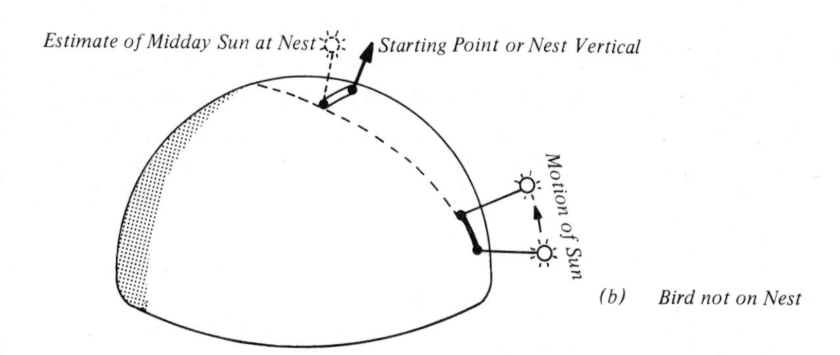

(b) Bird not on Nest

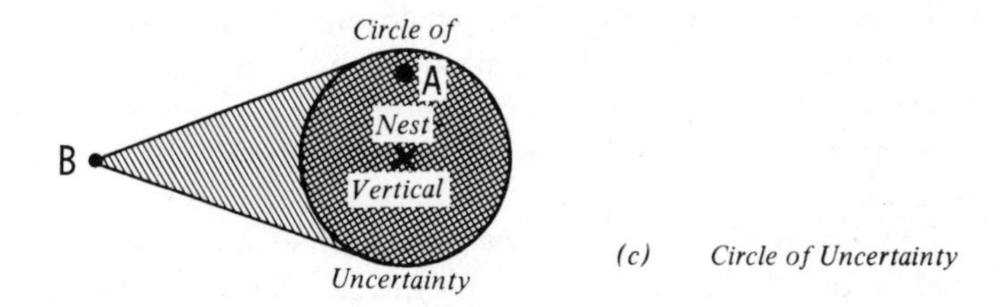

(c) Circle of Uncertainty

33. Navigating by the Sun

visual correction at the last moment. A bird homing to its nest vertical will have a similar problem and will have to complete the journey by visual landmarks, an easy problem for a Manx shearwater used to flying round the southern coasts of Wales, England and Ireland.

However, there is a special problem here. When not too close to somebody who hits a ball straight up into the air, it is easy to set out towards the place where it will land. However, if standing directly under the ball, it may be quite awkward and one may not know which way to face. A similar problem confuses a bird. Suppose the accuracy with which the nest vertical can be detected is 1°, which represents a distance of about 70 miles. If the bird is closer to its nest than this, it may not know which way to face to get to it.

In Fig. 33(c), the error in the vertical is shown as a cross-shaded 'circle of uncertainty' having a radius of 70 miles. If the bird is less than 70 miles from its nest, at a position such as A in the figure, it may, like the catcher underneath the high ball, be uncertain which way to go. On the other hand, if at B, 200 miles from the nest, it should be able to fly within the sector shown by the single shading. It is well known that pigeons, when unfamiliar with the ground around their lofts, take longer to home from 70 miles or less than they do when released much further away.[47] Also, if they get confused, they are apt to fly away from the loft to get their bearings. It will have been noted that the 70-mile circle of uncertainty agrees with a vertical error of 1°.

We shall now consider other sources of error. Suppose an animal's clock gains or loses time at the rate of 6 minutes a day, as suggested on page 37, which builds up to say 20 minutes in a week for reasons suggested on the same page. This would shift the nest vertical east or west by up to three hundred miles which suggests that either the shearwater had a very good timing system or else the long east–west coasts of southern Ireland and Cornwall and Devon helped to guide it to its nest.

The Sun itself also changes the time at which it reaches its highest point in the sky but this would shift the nest vertical east or west by less than 15 miles a day, a negligible amount compared to that caused by errors in the timing system of the animal. However, north–south errors due to the height of the Sun varying throughout the year would be much more serious. Except around midsummer and midwinter, the shift could represent over 200 miles in a week.

Although this remarkable feat of navigation may be subject to serious errors due to lapses in time or in date, in practice the Manx shearwater uses it to roam for a few days over the ocean foraging for fish, probably keeping an eye all the time on the direction of its home vertical. In this connection, Dr. Matthews has found that these birds do not set course for their nests except in the daytime, though they wait until it is dark before coming in to their burrows, for they fear predatory gulls.

The Sun could also be used to fly to an east–west line through the nest without any need to know time. It would be a matter of remembering the height of the 'midday' Sun at home and flying towards the path of the Sun if the 'midday' height at its actual position were too low or away if it were too high. When the actual 'midday' Sun was at its right height, it would be possible to home by flying parallel to the path of the Sun provided the bird was aware which way to turn to reach home, which could be obvious if it were foraging out in the ocean.

For migration, provided the course had not too much east or west in it, the bird could fly using the Sun for guidance until it reached the latitude of its destination where the local 'midday' Sun would coincide with the height of a 'midday' Sun imprinted on its brain. We shall find this is not beyond the bounds of possibility when we examine how birds navigate by using the stars, but it would depend on the bird migrating at around the right date which, in our discussion of yearly rhythms, we found within the capabilities of warblers even if kept in cages from which they could not see the Sun. In real life, they would be assisted by the length of the day.

Moon. By contrast, the Moon alters its time of closest approach to the nest by intervals of time which represent many hundreds of miles a day. Also the Moon changes its height above the horizon each month by roughly as much as the Sun does in a year. Therefore, although as we have seen the Moon might be used to find direction by estimating the highest point it reaches in the sky on any particular night, it could not be applied to finding latitude nor to indicating a northerly or southerly distance to travel to reach an east–west line through the home nest.

Stars. Dr. Franz Sauer and his wife undertook, amongst other remarkable experiments, tests with blackcap warblers reared singly from the egg in a planetarium.[48] These birds migrate by night south-east from North Germany until they reach the area of Cyprus and then, to avoid flying to their deaths in the Arabian desert, they turn south and travel to Kenya. Around the migration time in September, each fledgling would flutter to the south-east according to the stars in the planetarium. It did not matter if the planetarium were wrongly oriented nor if the night sky projected on the dome were largely occluded. Only if the planetarium were stopped from rotating were the birds flummoxed.

If the stars in the planetarium were then switched to show the paths they would follow in Prague or Budapest, the warblers continued to flutter south-east but, when the stars were rotated as over Cyprus, they fluttered due south. Imprinted on their tiny brains was probably the general rotational pattern of the night sky, the directions in which the stars rose and set vertically, the directions of motion of the stars at right angles to the north–south line, indeed the general pattern. Alternatively they might have recognised that the pole star was a third the way up from the horizon to the vertical and, though this seems simpler, we must remember that the birds were flummoxed when the rotation of the planetarium was stopped.

Finally, when the sky in the planetarium was set as for Kenya, the migration fever would depart and the birds would go to sleep. So their tiny minds must have registered either that the pattern of rotation of the stars was the same ahead of them as it was behind or, possibly, that the pole star had sunk down to the horizon. Truly a remarkable experiment!

It was also shown that the blackcaps used the same instincts for their return journeys. Birds kept in Germany throughout the winter had their return restlessness in March and, if shown the stars in a planetarium as they would be in Kenya, they fluttered northwards. The birds also had an instinctive idea how long their journeys would last. If kept indoors, they fluttered in September for the time it would take them to fly to Kenya and they fluttered again in March for the time it would take them to fly back to Germany.

We see that birds can navigate by the stars just as they can navigate by the Sun without using their sense of time. They find direction from the general pattern of star rotations across the night sky and use the same rotation pattern to sense the latitude. Yet it is inconceivable that birds should use the stars in the same way as the Manx shearwater uses the Sun, that is, by knowing the time at which a particular star will reach a point directly overhead. Not only would it be difficult for an animal to recognise a particular star but, even if it could, the nightly change in timing of the star would shift a position many tens of miles westward and, in any event, the star might not be up during the flight of the bird.

Yet the birds may well be able to remember star patterns for short periods. A remarkable experiment has been made with lesser whitethroats whose migration patterns are similar to those of blackcaps. The night sky in the planetarium holding the birds was suddenly switched 60° eastwards to somewhere in Siberia while the birds were fluttering south-east. The poor creatures looked shattered by this but they fluttered manfully to the west to put things right and, as the sky was shifted in steps back to its correct position, so their headings rotated through south to south-east. Their reaction is perhaps an additional indication that birds detect the highest points in the sky which stars and other objects will reach and can remember these highest points at least for a time.

OTHER METHODS

Dead reckoning. This means calculating a future position from a knowledge of course, speed and time spent at that speed. We saw in Chapter 2 how animals measure courses from Sun, Moon, stars and the Earth's magnetic field. However, land animals may find it difficult to follow steady courses. Their paths may be deflected by mountains, rivers or even by open spaces where they can be exposed unduly to hungry hunters.

Birds are better placed. They fly at reasonably steady speeds and many have an instinctive knowledge how long their journeys should last. As we have seen, birds exhibit restlessness if confined during their periods of migration. Willow-warblers who travel to South Africa spend longer periods of restlessness than wood-warblers who only go to North Africa. Blackcaps 'know' how long they have to fly to reach their migration destinations and this will act as a back-up check on their ability to use the stars to tell them when they arrive in Kenya.

Airmen use speed and time to find how far they can travel without refuelling. Birds may reverse this process and stock up instinctively with energy-giving foods before they start out. A small bird such as a sedge-warbler may eat until it doubles its weight before setting out on its migration. Each hour of flying can consume $\frac{1}{2}$ per cent of its weight but, as the bird will become lighter as the hours go by, it might well fly for 140 hours and travel for perhaps two thousand miles according to its cruising speed.

But let us not forget the amazing honey bee. When a worker returns from a promising nectar area and dances on the alighting board of a hive or on an upright honeycomb, the speed at which it dances tells the other bees how far away the nectar area must be. For, the longer the distance, the more tired the bee will have become and so the slower it will dance. This is sensed by the bees clustering round to find the direction of the area.

Magnetic navigation. So many animals have been shown to have an ability to steer by the Earth's magnetic field that we might expect them to navigate by it. It has been proved that birds at least sense the slope of the field and therefore can be expected to remember what it was at their homes. An animal may also notice that, when moving in the direction of the downwards dip, it steepens and, when moving the other way, it flattens. So the animal, by noting the dip, may be able to decide whether to travel towards the downward alignment, or to move in the opposite direction, to restore dip to its 'home' value.

This sounds reasonably simple and it is known that pigeons, released in the Kyffhauser mountains, where the dip is distorted due to local iron-ore deposits underground, travel at first in the direction opposite to their lofts. Furthermore, on the average, dip changes 1° for every 70 miles of north and south travel, which places it on the same sort of circle of uncertainty as any other system using the vertical. Without its vertical, an animal could not measure dip at all.

Fig. 34 shows how the dip of the Earth's magnetic field, marked with solid lines running roughly east–west, varies over the surface of the Earth, the lines being drawn for every 20° of change in dip. The north–south distance between any two points on the map is according to the constant scale shown below it. Thus, by travelling in a certain direction until the dip reaches a specific value, an animal could be directed instinctively to a given east–west line, just as it could be directed to a given latitude by using the Sun by day or the stars by night and without a knowledge of precise time.

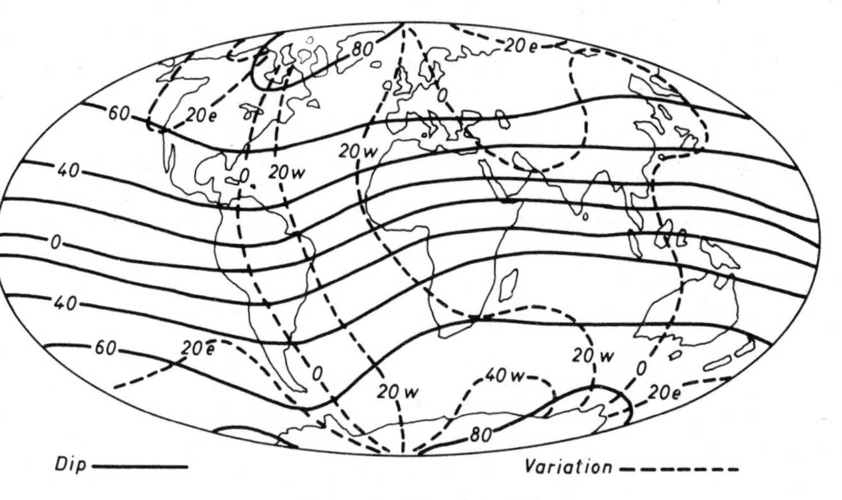

34. Magnetic dip and variation

It has been suggested that animals may find direction from the Sun, the Moon and the general rotation of the stars by using these objects to give a visual representation of magnetic direction, the astronomical and the magnetic alignments being generally close together. It follows that animals should be able to recognise the differences in direction between these two sources of information and this difference is known as 'variation'. It is plotted on Fig. 34, again for every 20°, and shown by dotted lines. Once again, the east–west distance between any two points is according to the constant scale shown on the map.

It is obvious that variation changes in a most erratic fashion, unlike dip. It is virtually unusable where the dip is more than 80°, being affected too greatly by magnetic anomalies and varying unduly in magnetic storms, the latter affecting all arctic and antarctic regions. Thus the only area where variation might be of use is in other areas where it changes at about the same rate as dip, that is, around the east coast of North America. There are signs it may be used in this part of the world.

It is known that, in north-east America, pigeons become upset and have difficulty in finding their homes where variation changes erratically due to local magnetic deposits underground. However, in other parts of the world, the problem of magnetic navigation is a repeat of the difficulties with Sun or stars; it is relatively easy to find position in terms of north or south latitude but to establish east–west longitude is another matter. So we shall not suggest variation as a method used by animals except in the one particular area.

Visual. The eyes are used to recognise landmarks, particularly those around an animal's home. Thus the young Australian aborigine would go on a 'walkabout' and learn, as he roamed further and further afield, always to know the exact direction in which his home lay. Even in these days, people may walk round an area in which they have newly come to live in order to get to know it.

It is certain that birds and mammals are able to identify objects they have seen even in the quite distant past. Pigeons taught to peck at a certain place in a photograph will repeat the process six years later. Mammals will not only recognise each other but also individual human beings, and birds have a similar ability. Sometimes experimenters need to disguise themselves when dealing with quite small animals but, on the other hand, one man who failed to establish the usual rapport with a starling discovered it was because his false teeth were undergoing repair.

We can therefore safely assert that mammals and birds have good pattern recognition and, in this respect, horses are remarkable. An Arab steed will recognise its master, among other men similarly clothed, a quarter of a mile away. Nor would anybody bother to conduct a homing experiment carrying a cat or a dog away in a motor car without blanketing the windows. In the dark it will be more difficult but there may be distinctive lights on roads, railways, houses and blocks of flats, while towns glow in the distance. Also, coastlines, rivers and lakes show up well even in starlight.

By comparison, reptiles, amphibians and fish may be poor visual navigators and, in many instances, their brains will be attuned only to things edible or dangerous and probably they have only limited abilities to recognise their surroundings by sight. In any event, many of them have no homes as such and therefore are less interested in recognising places. Insects are similarly somewhat stereotyped in their visual abilities and only recognise simple patterns, which inevitably may be repeated continually in one locality. Bees, however, in their early flights, are believed to learn the areas around their hives, but whether or not sight is the main factor is not easy to prove.

Because the essence of recognition is the arrangement of features topographical as well as facial, it is to be expected that animals need to avoid mistakes, due to common combinations and dispositions of ordinary things, and will pay particular attention to places which have a striking or unusual appearance. This habit is noticeable among pilots of light aircraft, who tend to look for 'sure-fire' landmarks and ignore the commonplace. It would be surprising if animals did not adopt similar unmistakable patterns to help them find their ways home.

There is, however, one completely different use of the eyes and that is, not for recognition, but rather for guidance. A bird in particular may follow a coastline, river or motorway, knowing it will lead to a point recognisable with certainty, from which the way home can be followed from memory. In this use of visual guides, it is likely that animals follow the practice of human navigators and 'aim off' to one side so that, on reaching the line they have to follow, they know which way to turn.

Not only are coastlines, mountain ranges and rivers valuable guides, but also there are lines in the oceans where water changes colour. In polar regions, there are large numbers of organisms in the water such as krill, and a reduced salt content due to the molten ice. As a result, the water looks very green and may be contrasted with currents of normal colour travelling the opposite way. At the other end of the scale, the water in the Sargasso Sea is extremely blue and the line of demarcation between the blue emerging water of the Gulf Stream and the normal colour of the Atlantic is also visible.

Seabirds, in particular the frigate-bird which becomes waterlogged if it settles on the waves, need to be able to detect the presence of land from a distance. They may be able to see 'standing' clouds over islands or over high ground which is below the horizon. Or they may detect it by the presence of other birds, particularly their flights from the land in the early morning and back towards it as the Sun goes down. Debris in the water is also an indication and the silt carried by fresh water from estuaries gives a greenish tinge to the sea which, in the instance of the River Plate, is visible a thousand miles off the coast of South America.

Vibrations. We have seen the wide use that fish make of their lateral lines. Naturally, they will detect the change in flow of water over their heads due to an obstruction. Thus a pet goldfish, not the most intelligent of creatures, never bumps into the glass sides of its bowl. So it is possible that fish use their lateral lines to find their ways about and they may even be able to identify a place underwater by the pattern in which the vibrations set up by their swimming is reflected. One suspects this might particularly apply to fish who live round coral reefs.

Sound. Animals in their travels may recognise sounds such as waterfalls, rapids and surf on beaches or sandbars, together with man-made sounds from factories, roads and railways. Pigeons and probably other birds can home onto the low notes coming from such sources by altering course until the waves arrive most frequently at their ears and therefore sound higher than when flying in other directions. This is another example of the use of the 'doppler' effect.

Although very high frequency sound waves are used for hunting, lower notes seem to predominate for finding the way. Low sounds even emerge from major topographical features, such as certain mountains, and act as acoustic landmarks. The baleen whale, living on layers of passive plankton and not possessing the hunting sonar, can use low frequency sounds to tell the depth of the water below and to find horizontal distances to shallows and rocky coastlines.

In a Texas cave, there may be twenty thousand bats, many of them flying about in pitch darkness, and yet they never collide. Each knows its own signature tune, the frequency and the changes of frequency, and can identify its own echo, though it must be rather like listening to a whisper at a Cup Final. Yet there is a curious feature of bats in caverns. Just as humans learn to find their ways at night about their bedrooms and may not bother to switch on the light so, once a bat is familiar with a cave, it may not feel a need to use its sonar. If an obstruction be placed unexpectedly somewhere, bats will bump into it until they learn where it is!

That this is due to overconfidence rather than to defects in sonar was proved by D. Griffin and R. Galambos of Harvard. They set up a curtain of thin wires a foot apart, through which they found bats would fly, only occasionally touching them. Even when the wires were only a third of a millimetre thick, the bats seldom hit them, not even when bombarded by supersonic noises from loudspeakers which were a couple of thousand times as loud as the echoes. The wires had to be thinned down until fourteen laid side by side made up a millimetre before they could not be avoided.

The large fruit-eating bats, often known as flying foxes, which naturally have no systems comparable to those that hunt insects at night, apparently retain a crude form of sonar. They make clicking noises with their tongues on wavelengths three times as long as the hunting bats' and, by listening to the echoes, seem able to avoid flying into trees at night when they feed.

Flying squirrels, which glide but do not fly, probably have some crude form of sonar for they too are nocturnal. Shrews can distinguish obstacles by echoes from sounds they make on a wavelength five times as long as a bat's. Many other animals must use sound in this way, particularly if they lose the use of another sense, and this applies notably to blind humans, who find their ways remarkably well by listening to the echoes from the tappings of their sticks.

It is not surprising that sonar is limited mainly to mammals for, apart from certain subtleties in transmission, very acute hearing with directional properties is needed. However, some birds may use it in a modified form. The Manx shearwater, returning to the burrow after foraging in the ocean, comes in to land at night to avoid predatory gulls and screams as it does so, even though its mate may not be at home, the scream stopping immediately on touchdown. The reflections of these screams from the surrounding terrain, to which the bird has probably become accustomed, seem to help guide it to the right burrow.

Smell. To the sensitive nose of an animal, every place has its characteristic scent and even a human being lost in a town may be aided by the odour of a local brewery. Animals, with their extremely discriminating sense of smell, use their noses to identify places and to support recognition by other means. This applies not only to mammals but also to birds, who certainly use the sense to help them return to their nests.

The scent of rosemary may be detected by people on board a ship thirty miles off the coast of Spain, and Australia was smelt by explorers before it was seen. This suggests the sense of smell is of particular value when there is not much conflicting or supporting information on which the animal can concentrate. Thus in World War II a well-known desert warrior was able to detect a camel at a distance

35. The leaping salmon

of six miles amongst arid sandhills, perhaps as a result of its pungency. The isolated Spice Islands were so named by the sailors who smelt them at a distance and indeed many mariners are able to 'smell' land. There can of course be no doubt that seabirds can do the same even further away for they will be free from local distracting scents.

However, it is fish that make the most striking use of smell for navigation. The salmon is a well known example. As pictured in Fig.

35, it fights its way upstream, leaping wildly up waterfalls and rapids, six foot or more out of the water, until it reaches the clear stream, rich in oxygen, where it was born and there the eggs are laid. Dr. W. A. Clemens of Canada tagged half a million young fry in the upper waters of a river and later caught ten thousand of them as adults all in the same tributary and none in any others.[49] Dr. G. H. Allen and Dr. L. R. Donaldson carried salmon spawn from one river to a stream in another and tagged them before they migrated. The survivors all returned to the stream in which they had been tagged and none to any other stream, not even that in which they were born.

Dr. A. D. Hasler finally proved that the ability to reach the stream in which a young fry was reared depended on a sense of smell imprinted on the minuscule brain.[50] He collected three hundred salmon on their way upstream to spawn, tagged them all and temporarily blocked the nostrils of half of them. All the fish swam upstream to the point at which the river divided but those that were unable to smell then swam about aimlessly, whereas the others continued to their allotted streams upriver. The magic had been demonstrated.

It has since been shown that the sensitivity of a salmon to scent is so selective that it will be put off by the paw of a bear dipped into the water upstream, but not by a solution of Brasso![51] Electrical tests of the olfactory lobe have revealed that, although the waters of the upper stream produce the strongest reaction, there are lesser responses to the scent in the river leading to the stream, the estuary of the river and even the open sea close by. The salmon is, of course, not the only fish with a remarkable sense of smell. A tiny eel is said to be able to detect a pin's head of chemical in a cubic mile of water.

Many animals, including ants, leave trails of scent when they travel in order to help them return to their nests. The barnacle ensures a safe return to its rock by a thin smear of a protein indissoluble in water, which is sensed by a receptor responding only to that particular protein. Hermit crabs detect dead molluscs by smell when searching for new accommodation. At the other end of the scale, mammals may mark their lairs and the boundaries of their territories with scent, though this is to discourage entry by rivals rather than to assist in finding the way home.

Summarising, it is safe to assert that smell is a major sense used to find the way. Even the toad on its travels of only a few miles can manage without sight, but not without scent. The female turtle relies on smell to bring her back over the oceans to the sort of beach where she was hatched. And it is the scent given off by neighbours, raised to unacceptable limits by overcrowding, that excites lemmings and insects, particularly locusts, to breed wildly and then set out on their emigrations.

Geographical. Apart from the ability of electric fishes to detect rocks, many remaining sources of information may be classed as geographical and naturally are not connected with hunting or avoiding. They include the funnelling due to mountain ranges and defiles, which circumscribes the motions of many insects. Perhaps even more important are the effects of the prevailing winds which, over the sea, will alter the paths of birds directly and, by causing ocean currents, may largely control the travels of many fish.

In the tropics, the Sun overhead causes hot air to rise and to cool, producing torrential rain along a line known as the intertropical front. This is so named because air, sucked in from north and south, meets along this line, being warped towards the west by the rotation of the Earth, these reliable inward-blowing easterlies being known as 'trade winds'. North and south the opposite occurs and winds blow away from the equator and are warped to become westerlies. Thus there will be a clockwise circulation of winds over the northern oceans and an anti-clockwise rotation over the southern seas. In addition air from the cold poles drifts westwards and towards the equator, in a direction opposite to the winds in the temperate zones.

A grossly simplified diagram of the prevailing winds, which shift as the Sun moves north and south during the year, is given in Fig. 36. The winds, like the migrations of animals, tend to be deflected by land masses, particularly those bordered by mountain ranges. Also the strength of the wind is affected greatly by what sailors call the 'fetch', the distance over which it has had a clear run.

As a result of these winds, the surface ocean currents follow a similar pattern, apart from a balancing equatorial counter-current travelling towards the east along the line of the equatorial front. Naturally, where the wind blows away from the equator, they will drift warm water with them and, where they move in towards the equator or emerge from the polar areas, they will tend to bring colder water. These flows of warm and cold water are of great importance to

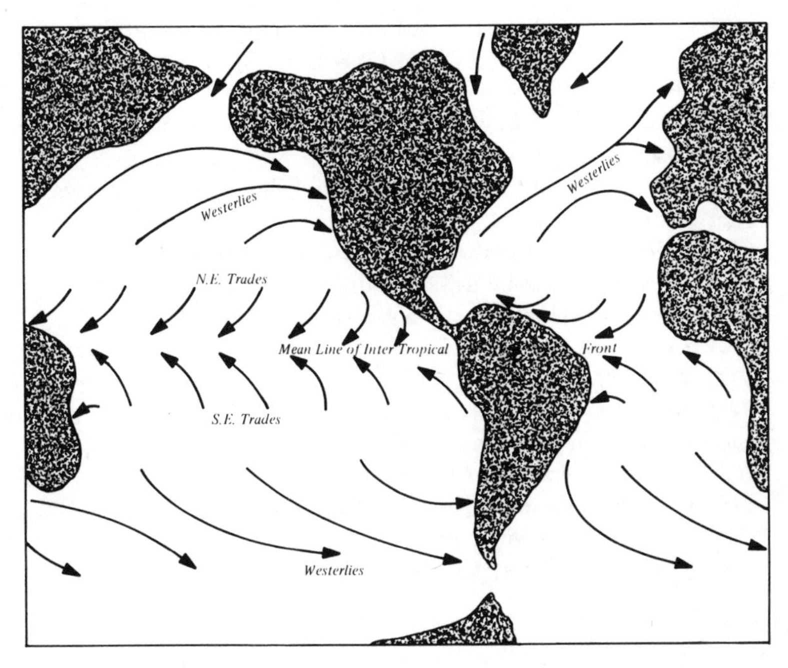

36. Notional pattern of prevailing winds

fish whose travels, like their very lives, often depend on suitable water temperatures.

Sundries. The winds also produce waves in the sea and, in particular, set up ocean swells which travel for long distances and, as we have seen, these may be used to indicate direction. The swells bend as they approach an island and meet again along a line of turbulence after passing by. In certain instances, minute luminescent sea creatures are stirred up from below and may form phosphorescent runways pointing towards the land. In addition, many ocean currents and also rivers have waters with an individual taste, including a particular amount or lack of salt, and this may be identified by an animal.

SUMMARY

To find the way, the following methods are known to be used, though there may be others:

1. *Visual** particularly by birds and mammals. Used in two ways:
   (a) *Position finding* by recognition or
   (b) *Guidance*, for example coastlines, roads and railways.
2. *Smell** particularly by fish and also by mammals and birds.
3. *Direction*, for which time is not essential. Using:
   (a) *Sun*,
   (b) *Moon*,
   (c) *Stars* or
   (d) *Magnetism*, detecting the slope of the Earth's field.

And *Distance* travelled along the direction. Found:
   (a) *Directly* from dancings of bees or using
   (b) *Indirect* means such as
      (i) *Time* of flight, sensed by birds or
      (ii) *Food consumption* by birds.

Or '*Latitude*' for which time is not essential. Using:
   (a) *Sun,*
   (b) *Stars,*
   (c) *Magnetic dip*, the slope of the Earth's field.

4. '*Longitude*' by magnetic variation, probably restricted to north-eastern parts of America.
5. *Position* by homing, using the Sun and relying on an accurate sense of time. Period since leaving home is generally limited.
6. *Electricity** only by fish producing their own fields.
7. *Geographical* using
   (a) *Topography*,
   (b) *Winds*,
   (c) *Ocean currents*,
   (d) *Temperature* of surroundings and
   (e) *Colour, smell and taste* of water.

* Used also for hunting and avoiding.

7

Journeyings of Cold-Blooded Creatures

Invertebrates, including primitive animals, crustaceans and insects, together with fish, amphibians and reptiles, take their temperatures from their surroundings. If they get very cold or begin to dry up in the heat, they may manage to slow down the rate at which they live which enables them to recover from extreme conditions, so that they can populate areas not available to creatures unable to reduce the rate at which they consume their reserves of energy.

Very simple single-celled creatures can withstand freezing and an extraordinary degree of desiccation. More complex animals die if their blood solidifies despite the antifreeze they produce and so, in winter, they tend to take shelter and dig down into the ground. Yet fragile butterflies may survive three months under the snow and the eggs of certain insects may even need a cold snap before they will emerge from their winter sleep!

Because they live in water, fish have less severe problems and generally move about all the year round, but they do not survive in lakes frozen solid. Naturally they die if the pond in which they live should dry up, though lung-fish burrow into the mud and, using an air bladder to breathe air, are able to recover. Amphibians have similar problems and so do reptiles, though to a lesser extent. They can withstand very low temperatures but die quickly in the extreme heat. Therefore reptiles in particular take shelter by digging into the ground. For this reason they do not live in areas where the ground becomes frozen. Snakes are an exception for they can wriggle into niches and have been found 15,000 feet up in the Himalayas but not in the Arctic nor the Antarctic.

Naturally fish will follow water temperatures to their liking and other cold-blooded creatures may journey to avoid the heat of summer or the cold of winter. These will be flying insects rather than animals that crawl about, such as land reptiles and beetles. Cold-blooded creatures also travel to areas where food is more plentiful and, in particular, to places suitable for breeding.

The reader will notice that details provided for bees, salmon and, to a lesser extent, for turtles are particularly full. This in part reflects the commercial importance of these animals and, in the instance of bees, it is supported by the wonderful work of Karl von Frisch. Yet none of these creatures has been studied by virtue of their navigational abilities and they should perhaps be regarded as examples of what flying insects, fish and swimming reptiles may, under certain conditions, achieve.

Zoologists often comment on the ability of an animal to find the way home when displaced to somewhere it does not know. The tests made with cold-blooded creatures generally involve transporting them a short distance away in covered cages to see which way they try to go. Snails are said to be good at it when carried a few tens of feet away and edible snails seem to go somewhere in the right direction from fifty feet. Ants and other insects generally fail to home to their nests when displaced sideways but miss by this displacement. Amphibians have some ability but fish and reptiles seem to take little interest provided they are not uncomfortable. In any event, this is not a fair test of navigational ability, but it may reveal details of the methods they employ. Obviously, the ants that miss their target are using some form of orientation.

INVERTEBRATES

As a general rule, primitive animals do not make homes and only move about to find food, to mate, to dodge enemies and to avoid conditions which are hostile. Others are moved about by the forces of nature, particularly those which are so small that gravity has correspondingly little effect on them. Spiderlings, mites and other tiny insects have been collected in nets by aircraft flying at over three miles above the ground.

At sea, the minute creatures that make up the animal element of plankton undertake what might be called a daily vertical migration. They live at considerable depths by day but swim up to the upper layers of the water at dusk. Some may go down again if there is no Moon and come up later, but all return to the comparative safety of the depths as the Sun rises. Many of these creatures are small crustaceans and, on this layer of plankton, small fish up to the size of herrings will feed. On them, larger fish will prey and so there will be a general daily vertical migration among fishes as well.

In addition, there are annual migrations. Clam worms shelter in rock crevices in the winter but, in the summer, they join the plankton and reproduce amongst them. However, perhaps the best known migrants in primitive creatures are the crabs, which have been known to travel over a hundred miles along the sea bottom in their spawning journeys. For example, the females move into shallow water to mate and lay their eggs and then return to the sea, followed with little enthusiasm by the males who tend to be less adventurous.

Chinese freshwater crabs live for several years in rivers but migrate to estuaries to mate. The female, with eggs attached, may then travel out to sea for the winter, but will return to the shallow water in spring. The eggs hatch out and, after a year in the estuaries, the young migrate upstream and grow to maturity in fresh water. Other crabs that have learnt to live on land migrate to the sea to reproduce, the directions of motion being determined probably by the Sun assisted by the general slope of the land towards the shore. Yet they will climb over small obstacles and even fall down cliffs rather than deviate to one side or the other.

Insects. The social insects that live in colonies have to find their ways to and from their nests and so they must learn to navigate. We have seen how bees forage along bee-lines using the visual, ultra-violet or polarised light from the Sun. Wasps similarly find their way back to their nests; so do ants but they also lay scent trails, while termites probably rely for direction on magnetic information.

Non-social insects may lay eggs in the autumn and die, the next generation appearing in the spring. Young female greenfly then drift with the summer breezes but the males are less ambitious. So the ladies carry on with virgin births until the males catch up in the late summer, not with them but with their daughters. Other insects may live for too short a time to set up a home or, if they do, never stray far from it. Many lay their eggs, cover them up and leave them. Not so the spider, who looks after her brood but even she, if forced to leave her web, simply builds another somewhere else. Hence the need to find a specific point is of interest only to the social insects or to their solitary relatives such as the little leaf-cutter bee, which flies to its hole carrying a large piece of green leaf as if it were an aerial surfboard.

Emigration. Many insects travel in huge swarms when there are too many in one place for the local vegetation to provide sufficient food. The action seems to be triggered off by a scent that evaporates quickly so that it can only be smelt by others close by. The scent seems to require the recipient to breed and thus overcrowding begets more overcrowding. The insects eventually emigrate en masse, generally in a fixed direction which means that at least they do not return to the place from which they started. There are two hundred and fifty species of moth and butterfly that emigrate, including the painted lady that lives in California and often flies in great swarms into the desert to perish.

It can be questioned whether emigrations to destruction can have a real purpose and without it they can hardly be classed as navigations. However, the instinct to move on when local conditions become unbearable may greatly increase the chance of survival for the species, particularly if accompanied by overbreeding to offset the additional risks inherent in any journeyings. Only when the overcrowding is exceptionally severe and an unfortunate choice of direction leads the emigrants into even worse surroundings, or to their deaths in the sea or the desert, do these emigrations assume the character of mass suicides. For the insect cannot know in which direction to travel and generally follows the prevailing winds.

The most famous emigrant is the locust, whose depredations have been chronicled in the Bible. It is a pale green grasshopper which, when overcrowded, grows wings and becomes darker. Thus the eggs of a winged locust, if separated into two batches, will grow into dark, winged creatures if crowded together but into pale green grasshoppers if given plenty of room. In normal conditions, the first step in overcrowding is for young locusts known as hoppers, whose wings have not fully developed, to travel over the ground, sometimes in huge armies, engulfing villages, fording rivers on the bodies of their struggling leaders and crossing burning grass on the scorched remains of their companions. Finally they take to the wing in vast hordes.

When airborne, a swarm of desert locusts may be ten thousand million strong[52] in a block of air a mile wide, fifty miles long and a hundred feet deep, 'darkening the Sun', and they descend at intervals to strip the land of every leaf and blade of grass, eating even those that are injured or die. One famous swarm covered a thousand square miles and, when blown out to sea by a strong wind, the bodies washed up formed a bank four foot high and over fifty miles long.

While most insects emigrate at the heights at which they normally fly, which is seldom more than a hundred feet, locusts may travel at several thousands of feet. They can fly at six miles an hour but, at such heights, the winds generally travel much faster. Thus, in spite of stopping to feed, they may cover several thousand miles in a month. Indeed, locusts have been drifted as far north as Ireland.[53]

The swarms appear to fly in a general downwind direction which, in the tropics, should carry them towards the intertropical front where the intense rains produce the conditions needed for locust eggs. Yet, although following the winds, the insects travel on fixed headings; for swarms have been seen to cross, one just above the other, each going in a slightly different direction. The headings seem to be determined by the Sun by day and by the Moon at night.

Locust swarms are generally accompanied by solitary wasps who, as soon as the swarm lands, set to work, digging holes, laying eggs and paralysing locusts to provide living larders for their offspring, finally sealing up the holes. When the swarm departs, the wasps also take off, leaving behind unfinished holes and paralysed locusts. Thus they continue their journeyings with their apparently amenable hosts.

Migration. The regular, seasonal journeys of animals from a breeding area to a place where there is more food, and where the animals may even breed again before returning, is of course the process known as migration. The painted lady, already mentioned for her suicides in deserts, breeds along the northern fringes of the Sahara and, in March, patches of grass shimmer with millions of butterflies struggling to emerge and dry their wings. They then fly northwards away from the Sun at six miles an hour and cross the Mediterranean, occasionally in large swarms.

Painted ladies can also navigate single-handed and they have been seen to approach a ship alone from one side, fly up and over the deck and then continue on the same straight course the other side. By May, they will have reached the north of France and finally, after a journey of around two thousand miles, they arrive at their summer breeding grounds along a line through England, Denmark and northern

Germany. In the autumn, it is back to the desert again, a quite remarkable feat for such a fragile creature.

In North America, the migration of the monarch butterfly, large and beautifully coloured but nasty to taste, is most spectacular,[54] and typical routes are shown in Fig. 37. Towards the end of August, these insects leave the northern parts of the United States and southern Canada and fly southwards, travelling 80 miles a day for 2,000 miles. They join up into groups, presumably as the result of some communal scent, and finally build up into swarms covering, when they land, every bush and tree over an area a couple of hundred yards wide and a mile or so long. Finally, they settle in certain trees in Florida and California, presumably being attracted by local odours, and act as tourist attractions until the spring, when they depart individually for their northern breeding grounds. A similar butterfly migrates to and fro between the equator and an area north of the Amazon river.

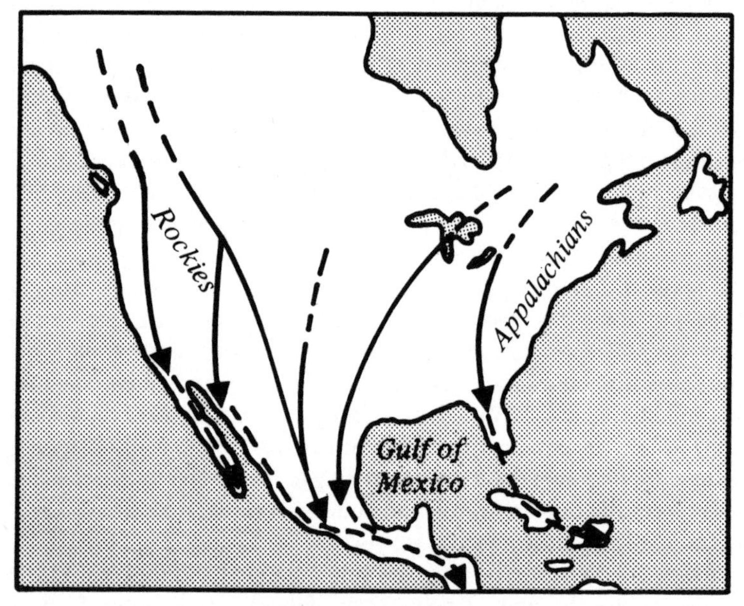

37. Migrations of the Monarch

On these daytime migrations, the butterflies probably rely partly on the Sun, their paths also being deflected by winds but constricted by the mountain ranges each side of the United States and split by the

Gulf of Mexico. There are similar remarkable migrations in other parts of the world. The cabbage-white butterfly crosses the North Sea from Scandinavia to England in July and August and it also negotiates the Alps, ascends the Pyrenees to get to Spain and even travels from Egypt to the Cape of Good Hope!

Navigationally, the death's-head hawk-moth, a creature that squeaks when molested, is even more remarkable for it flies by night, starting in North Africa and southern Europe and ending up in Scandinavia and England, and it has even reached Iceland, though one might suspect this was a combination of steering errors and unusual winds. It is extraordinary that an animal with such a tiny brain should be able to use the stars and even less likely it could make effective use of the Moon. Perhaps it has learnt to keep the Moon abeam when it rises and alters course thereafter by 45° every three hours or in some other way, following the pattern of the simple-minded dung-beetle.

Keeping warm. Being kept at the right temperature only by their surroundings, social insects huddle together in the winter, sometimes digging down into the ground. As it becomes colder, their activity decreases and eventually they may go rigid. When spring comes, a few who have managed by luck or good judgement to retain some ability to move will venture forth, become warmed by the Sun and return to give sufficient heat to their fellows for more to follow their example. So the whole colony recovers and goes to work once more.

Non-social insects may make journeys to follow the Sun. Thus swarms of dragonflies travel northwards and southwards and have even arrived in southern Ireland after what must have been a five-hundred-mile flight from Spain. Domestic fowl naturally welcome these edible visitors but unfortunately they may carry a worm which affects the birds and prevents them laying. So when a swarm appears, the local people shut their chickens and turkeys up in sheds or take them into their houses.

Other insects simply become dormant, generally huddling together to keep out the worst of the cold. American lady-birds fly to the lower slopes of the mountains for their winter sleep, returning to the valleys in the spring to devour greenfly. To help them, experts collected them in winter and released them in spring among the crops but they all departed elsewhere, knowing instinctively they had to fly to the feeding grounds and not having been informed they had already arrived.

Keeping cool. Not only do the lady-birds need to keep warm in winter, but also they have to keep cool in summer and so, those that live in California and in other places where the summer is hot, migrate to the higher slopes of the mountains in June and remain dormant until the autumn, and as many as thirty million may congregate in a quarter of an acre.

The Australian bugong moth also flies uphill at night from the plains of Victoria into the mountains to avoid the summer heat and a thousand may cluster on a couple of square feet of cave surface. Waiting for them will be parasitic worms which develop in the moths' bodies but are careful to depart before the moths fly away. The worms then mate, and lay eggs in winter so that their young can welcome the returning moths in the summer, a return presumably ensured by the funnelling effects of the valleys and the familiar odours of the particular caves.

The army-cutworm moth, so named because its caterpillars move about in hordes cutting off young crops at ground level, undertakes a similar 'summer hibernation' in caves 6,000 feet up in the Rocky Mountains. There it is welcomed by a predator somewhat larger than the parasitic worm, namely the grizzly bear.[55] A pestilent plant bug of the near East which fattens on growing barley flies 8,000 feet up into the mountains in summer and then enjoys a winter hibernation at 5,000 feet before returning to the barley fields in the spring.

Meanwhile, to keep cool in summer, social insects with wings fan the air in their homes and termites build additional ventilators. Other insects without permanent homes simply fly uphill if they feel too hot and downhill if it turns cold. Thus, in summer, the little long-tailed blue butterfly has been seen at 7,000 feet in the Pyrenees and even higher in the Himalayas where the resourceful hoverfly has been found at 12,000 feet.

FISH

Like many insects, fish do not generally operate from nests or lairs but, if they do, seldom depart far from them and return assisted by sight, smell and the reactions to surrounding objects of lateral lines, which act almost as a form of radar. In many instances their journeys are linked to water temperature and to the food on which they live. In connection with the latter, one should mention the remarkable little goby fish which lives on animals stranded in pools of seawater at high tide. It leaps from one pool to another although it cannot see where its target is but it never misses, having carefully surveyed the ground beforehand when the tide was in!

Like insects, fish also migrate, most of the more spectacular journeys depending on their breeding habits. These take one of three forms:

(a) Freshwater fish that migrate to the sea to breed,
(b) Seawater fish that migrate to fresh water to breed and
(c) Fish that both live and breed in fresh or salt water.

Freshwater fish that breed in the sea. The most celebrated example is the eel, found in waters all over the world, but we shall only consider those that breed in the Atlantic in the Sargasso Sea. The elvers, transparent leaflike creatures about a quarter of an inch long, travel with the North Equatorial Current and the Gulf Stream up the East coast of the United States, as shown in Fig. 38.

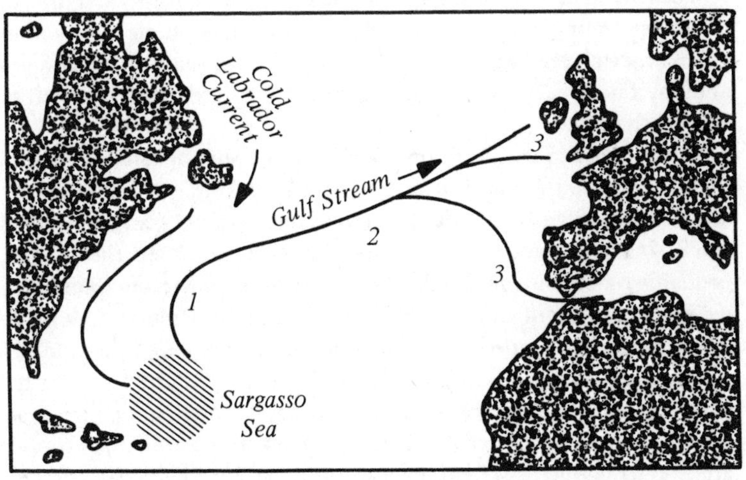

Note. Numbers on map suggest years after spawning

38. The drifting eels

Dr. Johannes Schmidt of Denmark,[56] has suggested that those on the western side of the Gulf Stream are deflected still further westwards into American estuaries, arriving about a year later. Those in the centre of the stream meet the cold Labrador Current and die. Those to the east differ from elvers on the west by having more bones in their spines, but this can be reproduced by hatching and rearing in certain temperatures. At the end of the year, these are about three quarters of an inch long, feeding voraciously about fifty feet below the surface of the Atlantic. Three years out from the Sargasso Sea, such as have survived predators are about three inches long and have drifted to various European estuaries, even entering the Mediterranean Sea. The American eels have meanwhile spent two years growing.

So, more by luck than by good judgement, the eels arrive at their estuaries, the larger the number that arrive, the more predators are waiting for them. The smaller and more concealed the estuaries, the fewer eels arrive but the less likely they are to be eaten in the process. Once in fresh water, the eels instinctively struggle against any current and fight their ways upstream, even wriggling uphill across fields on rainy nights. So the European and American rivers, land-locked lakes and ponds are replenished.

In the next five years or so, the eel grows into a powerful animal, females being up to 4½ feet long but males only half that size. Towards the end of their period in fresh water, they change colour from yellow to silver, cease feeding and travel downhill, once again crossing fields at night if necessary. After a short time getting used to salt water, they take to the sea.

The American eel, with only a relatively short distance to travel, then returns to the Sargasso Sea to breed, presumably following the sea bottom down to the depths and aided by a very acute sense of smell. It is very interesting to note that eels can orient themselves, probably by the Sun and, if displaced fifty miles or so from their home waters, are generally able to return to them, it apparently making no difference whether their sense of smell is temporarily inhibited.

But what of the European cousins? Even Dr. Schmidt cannot tell what happens to them for none has ever been found on the high seas. Theories that they can make their ways back to the breeding grounds are also discounted. The distances are great and the water currents unfavourable unless the fish follow the deep countercurrent below the Gulf Stream and this would mean a prodigious journey without any food. It is known that, at one time, Europe and America were much closer together and the European eels could then return to the Sargasso Sea. But now the continents have drifted further apart and it seems that the magnificent fish just go back to the sea to die.

Saltwater fish that breed in rivers. The life story of the salmon is well documented. The eggs are laid in late autumn, buried in the gravelly bottoms of fast-running streams high in the hills. In spring, tiny transparent creatures hatch out, each carrying a yolk sac for sustenance. By the end of the summer, the fish are two or three inches long and, at the end of two years, have grown to about six inches. Having grown for nearly three or four years, they generally drift gently down to the estuary and, after a pause to become acclimatised, they swim out to sea. Although this is a typical pattern, there are great variations. Some go to the sea at the end of the year after they have hatched, while others stay in the rivers until the seventh spring.

The young salmon, swimming out into the ocean in the early spring, will now be dark blue above and silver underneath and it lives and grows in the sea for one, two, three or four years during which period it may swim for a considerable distance. Salmon from the Pacific coast of North America head NNW or W and, after a couple of years of travel, during which they may cover several thousands of miles, they return to their estuaries to spawn in the rivers and die. Those from the west coast of Alaska head SW and have been caught as far away as South Korea, a distance of 3,500 miles. On the other hand, salmon in the Yukon river, which is 1,250 miles long, never come out into the sea and live all their lives in fresh water.

European salmon may go back to the sea and then return to spawn a second time. Those from Baltic rivers may stay in the Baltic or go into the North Sea. Scottish and Swedish salmon may range anywhere between Greenland and Spitzbergen, one from Sweden being found no less than 2,500 miles away. However, although the way that salmon navigate up the rivers to their breeding ground is well established, how they return to their estuaries is not known.

We know from tests of the olfactory lobes of these fish that they can sense the water round their estuaries[57] and it is also believed they can distinguish water by taste, as some human beings are able to do. Experiments on the shores of Lake Mendota strongly suggest that these fish can record the angle which the Sun reaches at midday, distorted as it is by the bending of the light at the surface of the water. Certainly salmon remember the scent of their 'home' estuary and so they might also remember the height of the midday Sun, albeit as seen from a fish's eye point of view.

Salmon travel from their estuaries generally east or west and, as they are known to be able to orient themselves by the Sun, they could be expected to return to the coast they left by reversing this motion. On arrival at the coast soon after midsummer, they would find the Sun higher than it was when they left the estuary in spring. Knowing that the nearer a cliff is approached, the higher it appears to be, the salmon could swim away from the Sun until it was at the remembered height.

In late summer, the midday Sun would be higher than in the spring but it would gradually become lower, leading the salmon southwards until it was at its 'home' estuary. This would fix the date of arrival at the estuary as being the same number of days after midsummer as the estuary was left by the salmon before midsummer. If, as seems likely, the salmon all left this particular estuary as soon as the temperature of the sea water was satisfactory, it would mean they would all return to breed at about the same date, male fish as well as female.

Although such an astronomical system might be postulated and could have certain advantages, the method of navigation might be magnetic or it could simply be a matter of drifting on the ocean currents down to the estuary of departure. In late summer and autumn, sea water flows southwards down the west coast of America and, in the absence of the well-remembered smell, a salmon might be expected to swim with it until it sensed the estuary, upon which it would swim against the flow. Nevertheless, it is interesting to note that plots of salmon in successive positions in the Pacific seem to support nicely the theory that, in the autumn, they follow the slowly descending Sun.

Trout are related to salmon but most of them spend their lives in rivers though a few go to sea. They too bury their eggs in fast-running gravelly streams and return to breed in the water where they were hatched, but perhaps with less certainty than salmon, which suggests they may follow other trout. Mullet and bass live in the sea, but compromise by breeding in coastal waters, grey mullet even travelling up rivers for short distances, and the shad, a species of small shark, does likewise.

The lamprey, from a surfeit of which that greedy opponent of the Magna Carta, King John, is said to have died, instead of a mouth has a sucker to extract blood. Like the salmon, it lives in the sea but spawns in rivers. So do most sturgeon, which hatch in rivers in the spring and go to sea at the end of the summer, though some remain in lakes and rivers all their lives. The females grow to a great size, as long as twenty feet, and travel with a retinue of smaller males as they parade around.

Fish that live and breed in salt water or in fresh water. We have already seen that water temperature is a major problem with fish, for their bodies are unable to compensate for it and, if they become cold, their vitality is decreased and their efficiency in catching other fish or in avoiding being caught is lowered. What may be even more crucial is that they need a certain water temperature to breed and often certain chemical conditions, including the right salinity if they live in the sea. To get the right surroundings and temperature, freshwater fish living in a lake will search out a suitable place for breeding and return to it year after year.

In the sea, fish may undertake yearly movements which follow changes of temperature due to the path of the Sun, moving to the north of the equator in the northern summer and to the south of the equator in the northern winter. The sole likes 20°C in March and April, a temperature only found in tidal waters. So, if the water is a little chilly, it swims with it and, because the colder water is due to the incoming tide, reaches warmer surroundings. If it feels a little warm, it swims with the water again and, as this must be an ebb tide, it eventually arrives in a colder area. In June, it prefers 14°C and so it repeats the same procedures to find water of this temperature.

In the oceans, sea temperature is greatly influenced by the circulation of the water due to the winds and this was discussed in the previous chapter. Fig. 39 shows the currents and their circulations, clockwise in the northern hemisphere and anti-clockwise in the southern seas, and also the warm and cold currents. One very important area is where the cold Labrador Current meets the warm Gulf Stream. The American shad likes temperatures between 13°C and 18°C and, in summer, it finds these in this area. In autumn, the fish travel 2,000 miles southwards against the Gulf Stream to winter not far from Florida.

The cod swims to spawning grounds on the north side of the Gulf

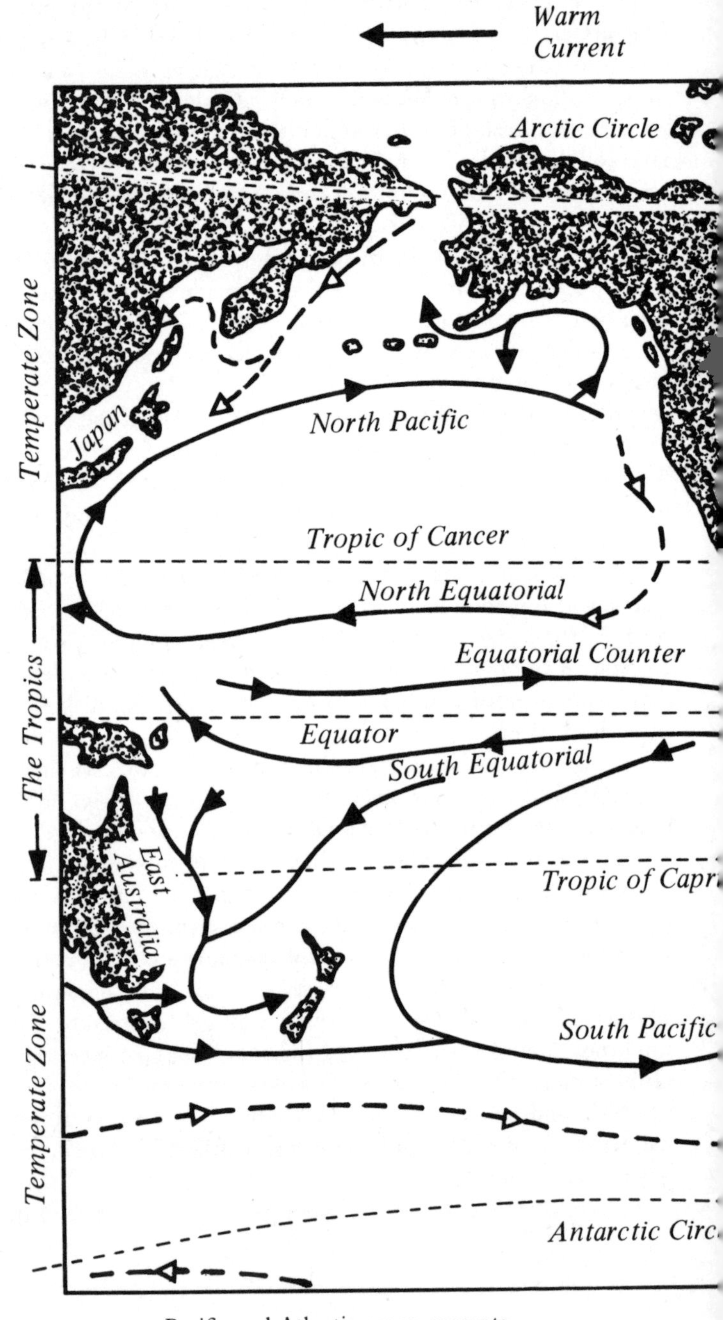

39. Pacific and Atlantic ocean currents

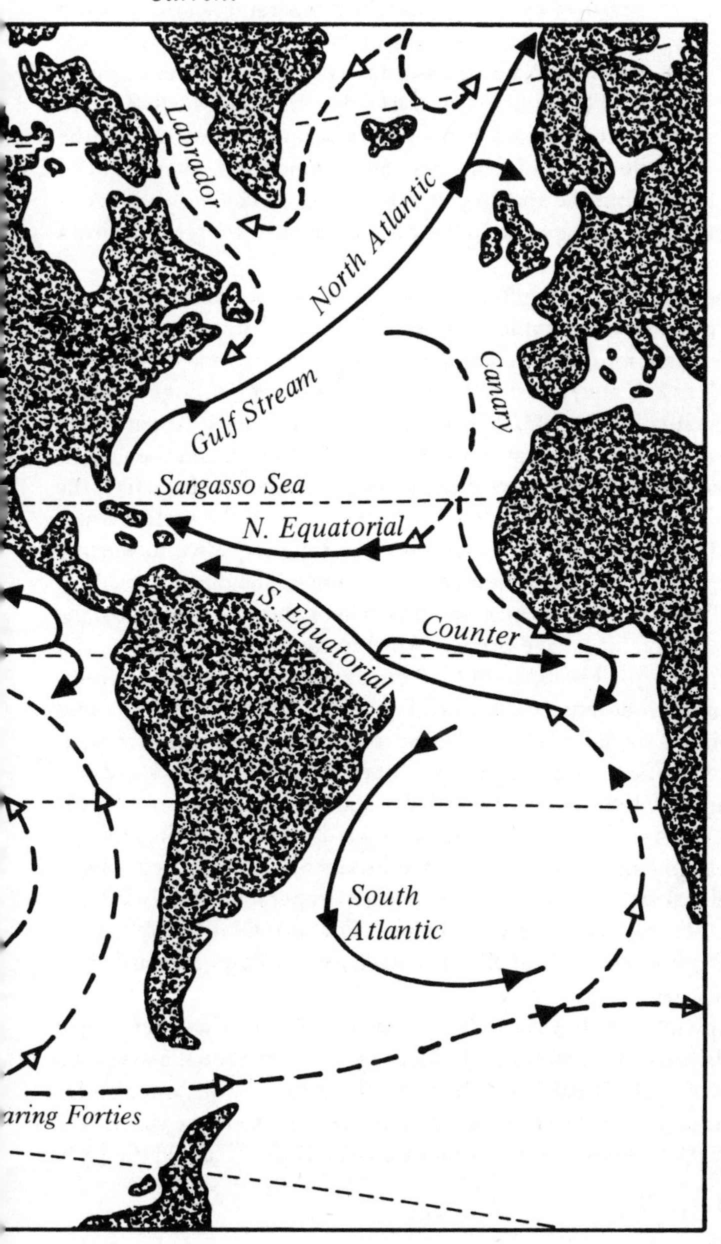

Cold Current
Labrador
North Atlantic
Gulf Stream
Canary
Sargasso Sea
N. Equatorial
S. Equatorial
Counter
South Atlantic
aring Forties

Stream where it becomes the North Atlantic Current flowing from Newfoundland past Iceland. The herring, which lives on plankton and does not eat fish of any size, is equally temperature-conscious and also likes a particular proportion of salt in the water, which it finds between the warm Gulf Stream and the colder and slower currents from the Arctic. These two types of fish also circle the North Sea and breed between the English Channel and the Norwegian coast. At the lower end of the range of fish sizes, the sardine likes water between 10°C and 17°C but the anchovy prefers the higher end of this range, a temperature it manages to find in the estuaries of large European rivers.

Some migrations are influenced by ocean currents and those of the tunny provide an example. These magnificent fish are not strictly cold-blooded, for they can maintain a temperature inside their bodies as much as 8°C above that of the water round them, though they are very sensitive to the cold. Those that hunt in the north Atlantic may spawn in spring in the warm water round the tropics and then travel northwards, following the requirement that, at a depth of 150 feet, the sea temperature must not fall below 14°C. Then, if the surface water gets cold at night due to evaporation, the fish dive down to warmer water below. By the beginning of June, those on the European side of the Atlantic reach the Spanish coast where they are joined by the bluefin tunny, who have bred in April in the Mediterranean.

From Fig. 39, it is apparent that, once clear of the Canary Current, the tunny can be expected to travel northwards towards Iceland and that some of them will round the British Isles and hunt for herring in the North Sea. Indeed, bluefin have reached Norway by the beginning of July, having travelled 3,000 miles in a month. As the summer ends, the tunny travel southwards again, bluefin presumably recognising the entrance to the Straits of Gibraltar as a unique estuary *into* which the sea flows, due to the evaporation of the Mediterranean. It has had to swim against this current to come out in the early summer and, in all other inlets, water flows outwards into the sea.

Meanwhile, the tunny on the continental shelf on the other side of the Atlantic travel from the tropics as far as Nova Scotia, assisted by the warm Gulf Stream and likewise fishing for herring and also for shad. In addition, tunny seem often to transfer from one side of the Atlantic to the other, perhaps following the Gulf Stream in the early summer and returning in the autumn by the North Equatorial Current.

In the Pacific Ocean, tunny are known as tuna. They make use of the strong clockwise current in northern waters and so their paths are relatively simple. They breed close to the tropics and go north as the Sun gets higher in the northern skies but are drifted westwards by the North Equatorial Current and so follow a curve along the coast of Asia. They are then carried north and eastwards and, as the Sun's path moves southwards, they turn south and complete the circle, going down the Californian coast and into the North Equatorial Current again, the six-year-old fish then moving further south towards the tropics to breed. One particular type of tunny, the bonito, hunts in packs and is so fast it sometimes gets airborne in pursuit of flying fish, though it cannot glide for a hundred yards using fins, as they do.

It will be noted that tunny work up and down the coasts of Europe, North America and the Far East because the relatively shallow continental shelves are full of prey. Spiny dogfish, small three-foot sharks, also roam along these areas in packs, hunting voraciously for herring and other fish, and moving up and down the coast in company with their prey. They also hunt mackerel which live far out over the continental shelf in March but move closer inshore by the end of June, spawning in the water and letting their heavy eggs sink down to the sea bed. Each shoal of mackerel makes for its own winter quarters at depths of 600 feet or more, the fish hiding in gullies or between sandbanks.

Swordfish also breed in tropical waters but then spread north and south, even entering arctic waters. Sharks are believed to spawn in shallow water near the equator before going out to hunt on the high seas. One can expect all fish will be influenced by ocean currents, and that they will follow the general flow patterns, relying particularly on the currents that drift over the continental shelves. Also, with their very acute sense of smell, they will be influenced by the special scents associated with their favoured breeding grounds. Yet fish may at times use the Sun to help them keep steady courses, for experiments with polarised light suggest that, by rotating polarisation, their orientations will often be altered.

AMPHIBIANS AND REPTILES

Amphibians. Because these creatures evolved from fish, their skins dry up very easily and so they tend to live close to water or under damp vegetation, while toads in particular hide during the day to prevent the evaporation of water due to the Sun. For most of the year, amphibians are solitary and do not move about a great deal. However, they undertake considerable journeys in the spring and congregate together in order to breed.

Newts make journeys from higher ground to specially chosen streams and, if displaced by a few miles, tend to return to their favourite waters,[58] where they stay for longer periods than some other amphibians. Frogs and toads also find their ways to a particular pond or lake, generally only travelling a few miles but sometimes crossing other ponds on the way. Finally, they collect together in large numbers to mate and spawn, after which they disperse to their normal living areas. So strong is their attachment to a favoured area of water for breeding that, if it should dry up or be emptied due to action by human beings, they may not reproduce at all. Furthermore, when a road has been built where their pond had been, toads have congregated hopefully on the site and have been run over in hundreds.

For navigation it seems these creatures travel uphill away from their breeding waters, forking out in various directions where the slope is not obvious but where there are chemical or other landmarks meaningful to them. The return journey seems to be a matter of going downhill and using these landmarks at crucial points of doubt. Towards the end of the journey, calls made by the males, for which frogs have resonant pouches in the throats, will give guidance for the last half mile or so.

It has been shown that amphibians may steer by the Sun, Moon and stars and that, when the sky is overcast, they appear to become disoriented. Perhaps they use objects in the sky to help them travel along reasonably straight paths, particularly over rough ground. Mostly they lay their eggs in the water and abandon them but some frogs lay them on land. In this event, the males may carry them into the water on their backs or hatch them in their resonating pouches while, in certain varieties, the females carry them about in sacs behind, like papooses.

Reptiles. With few exceptions, snakes, alligators and other reptiles are navigationally undistinguished. Indeed, the two types of animal specifically mentioned have only one fifteenth of 1 per cent of their body weight devoted to brain as compared to 1 per cent in a cat. However, to be quite fair, it has been suggested brain weight should be related to the area of an animal's body but, even on this basis, reptiles do not appear to possess great mental powers.

If the temperature around a reptile is too low, it becomes sluggish and cannot hunt, few of these animals being able to feed themselves in any other way. If they get too hot, they tend to die which makes them cautious about the Sun. If conditions are not to their liking, instead of setting out for somewhere else they seem to sink into lethargy. This is not altogether surprising for locomotion is generally hindered because their legs stick out of the sides of their bodies. No wonder reptiles tend to be solitary and collect together only to breed with others of the same ilk.

There are exceptions to this generally discouraging pattern. Flying lizards have developed membranes between 'fingers' and 'toes' and have skin flaps each side of the body and this enables them to glide from one tree to another. Giant tortoises live in humid areas but struggle slowly and cumbrously thirty miles or more downhill over rough terrain to lay their eggs in sand, the young emerging in due course and working their ways uphill to a climate more congenial.

The only real navigators are the turtles. Not so much the freshwater varieties, though these swim in large shoals in the breeding season and congregate on certain sandbanks in the rivers, but rather the sea turtles. These are known to travel fifteen hundred miles out into the ocean to certain islands to feed and they return to lay their eggs on the same sort of sandy beach as that on which they were born. Indeed, some say they return to the same beach, a remarkable feat for any animal and quite magical for a reptile. Probably, like fish, they rely on ocean currents, for bottles dropped into the sea off the beaches where they began life, end up among the islands where they feed, but they probably use scent carried on the water to find their way back[59] and may be assisted by an ability to orient by the Sun.

SUMMARY

The following are selections from the text:

1. *Invertebrates*
   - (a) *Plankton* migrate vertically each day followed by fish.
   - (b) *Crabs* breed in shallow sea.
   - (c) *Butterflies* may fly 4,000 miles yearly, generally north and south.
   - (d) *Moths* may travel at night and may go uphill in summer.
   - (e) *Locusts* may breed to excess and emigrate and many moths and butterflies do likewise.
2. *Fish*
   - (a) *Eels* breed in salt water, notably the Sargasso Sea, but grow in rivers and ponds.
   - (b) *Salmon* breed in mountain streams but grow at sea, travelling thousands of miles.
   - (c) Most fish live and breed either in
     - (i) *Fresh water*, breeding locally, or in
     - (ii) *Salt water* and may cover wide areas to feed and breed.

     Examples, tuna, shad, herring, cod and sharks.
3. *Amphibians.* Journey downhill to breed in ponds.
4. *Reptiles.* Travel little except for sea turtles which may go 3,000 miles to and from beaches where eggs are laid.

8

Journeys by Warm-Blooded Animals

By and large, birds and mammals are able to maintain their temperatures reasonably steady. They may spend part of the winter asleep in a sheltered place but only a few hibernate, which means a return to the cold-blooded state in order to conserve internal stores of energy. Birds, like aircraft, consume fuel at a high rate when they go about their business and need continuous supplies of food. So it is not surprising that only one of them hibernates, the American poor-will, known to the Red Indians as the sleeper. Humming-birds go into a hibernation when they sleep at night but, being so small, can emerge quickly from their dormant condition. Wrens also roost in communities in winter and become dormant but, again, they are small birds.

A similar situation arises with mammals. If they are large, they use so much energy in recovering from hibernation that it would not be economical. It has been estimated that the demands for stored energy to cope with arousals is ten times as great as that needed to maintain a hibernation and the time taken to enter into or come out of this state would be prohibitive. So only small mammals hibernate, namely hedgehogs, bats and a few small rodents, some of these animals generating potassium to act as a blood antifreeze.

Because warm-blooded animals keep their brains generally at the right temperatures and active, not only while they are awake but also when they sleep, their mental abilities are unimpaired, their learning processes are not interrupted and, in particular, their memories are developed to a much greater extent than those of their ancestors, the reptiles. As a result, we find their methods of navigation depend very largely on memory. Experiments involving carrying them away from their homes to places which they do not know therefore tend to give disappointing results, except with sea-birds and others that roam widely and need to be able to return home after their foragings.

Once again, it will be noticed that details regarding shearwaters, pigeons and bats are particularly complete. The shearwater has been intensively studied by Dr. G. V. T. Matthews, the pigeon is a particularly convenient bird being semi-domesticated and the bat has links between its sonar and modern radar. As before, we should not regard these three as necessarily outstanding in terms of navigational ability.

BIRDS

We may start by disposing of certain misconceptions. Egg-collectors in the past have claimed birds cannot count. Yet parrots have been taught to take up to seven pieces of food according to the number of light flashes shown to them or by the number of notes sounded on a flute, even taking two pieces when two notes were sounded together. Nor do birds lack colour discrimination, as we may infer by the plumage of cock-birds. A jackdaw has been taught to open a number of boxes with black lids until it has taken 2 peas, with green lids until 3 peas have been found, red lids until it has 4 peas and white lids until it has taken 5. Nor are birds' memories short. Pigeons will return to old lofts ten years after being moved to new premises.

Birds, as we have seen, have remarkably acute eyesight and will recognise shapes even when young. The ability to identify patterns is crucial in visual navigation. Pigeons have been trained by means of rewards to peck at human beings appearing on photographs, and could recognise them as such irrespective of clothes or absence of clothes and in spite of their colour. So we can assume birds will be adept at recognising landmarks and other objects in their surroundings. From the air, lakes, rivers and particularly coastlines are most distinctive and are generally visible even by starlight, a feature of great value to sea-birds.

Fig. 40 attempts to suggest how far a bird can see according to the height at which it is flying and the height of the ground at which it is looking. On the graph, the horizontal lines, in accordance with the left-hand vertical scale in feet, show the height of the bird. The vertical lines, with the scale along the bottom also in feet, allow for the height of the distant ground. The sloping lines evenly spaced represent the distance at which the object can be seen assuming the air is clear.

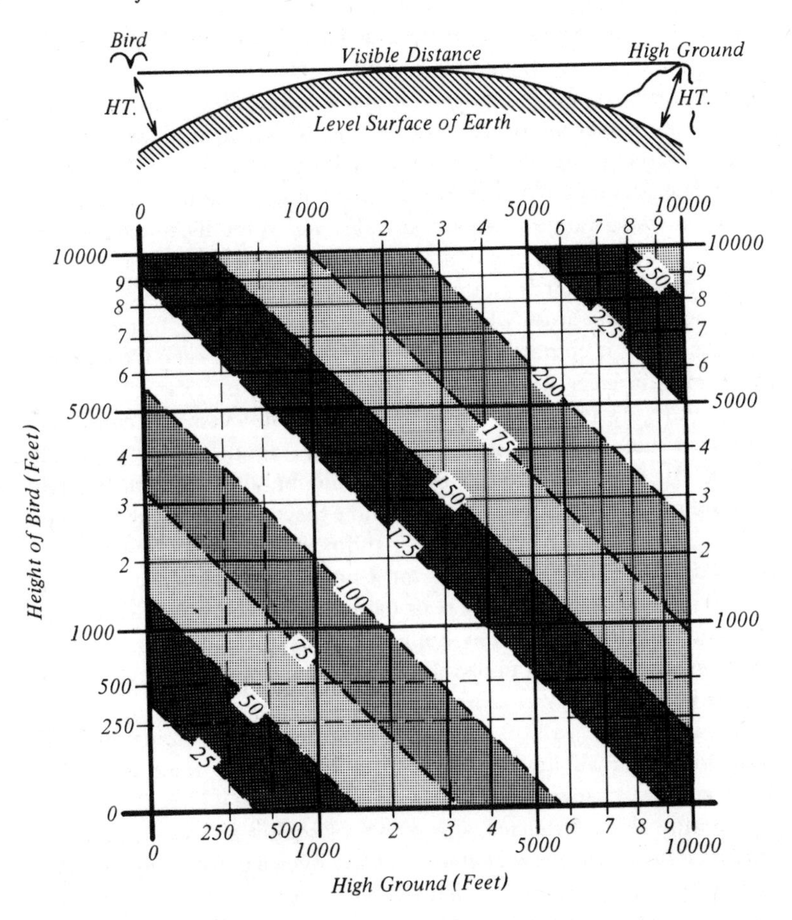

Numbers on Sloping Lines show Visible Distance in Miles

40. Height and visibility in clear air

In the cartoon above the diagram, the Earth is drawn as if its height were constant, as it would be on the sea or on a flat plain. However, it will be apparent that the only requirement is that, on the bird's horizon, nothing is higher than the ground below the bird and also that there is no shielding by high ground either side of the horizon. Thus, where the sloping lines cut the left-hand scale of the graph, indicates how far a bird can see over flat ground. Where the sloping lines cut the bottom scale, shows how far coastal high ground would be visible to a bird flying low over the surface of the sea.

Generally, birds fly higher than usual when they migrate. Small birds may travel as low as 200 feet, at which height they should generally see 20 miles all round or pick up ground 2,000 feet high when it is 60 miles away. However, many migrate at over 2,000 feet, giving a visual range of about 60 miles and some fly much higher, 10,000 feet not being uncommon, for quite small birds fly across the Himalayas.[60] Sea-birds generally fly low, perhaps on the look-out for fish, but, in the areas where they live, high ground of one or two thousand feet is common near the coast, and can be seen from about fifty miles in clear weather.

For homing to its nest, Fig. 40 shows that, even at a couple of hundred feet, a bird should be able to see about twenty miles all round. In reasonable visibility, it should be able to identify high ground much further away. One would expect that birds, like human pilots, work by means of 'sure-fire' landmarks from which they know how to fly to their nests. Thus for a bird with experience of flying round its neighbourhood at a reasonable height, there should be little difficulty in finding the way home from distances up to a hundred miles except at night or in bad weather. For sea-birds, these distances might be doubled even at night unless visibility is poor.

Yet experience will be important. Finding the way demands local knowledge and will depend on the distance round its home at which a bird habitually ranges and also the height at which it flies. Thus a tit, flying only short distances with plenty of shelter all round, may not find its way back if carried ten miles away. Even a pigeon, kept in a loft with a high fence round so that the surrounding country cannot be seen, may not be able to return home visually but will have to rely on a strong smell close by. Conversely, if a branch holding a nest be lopped off, the parents will fly in to where the nest was and, utterly bewildered by apparently missing it, will try again and again.

At night, or in the darkness of a cave, the final approach may be aided by a crude form of sonar and the screams of the Manx shearwater have already been mentioned. Cave swiftlets, from whose nests the famous oriental soup is made, and also oil-birds, navigate to their brood in caves and avoid bumping into each other by making clickings with their tongues and listening to the echoes.

Homing experiments have shown birds are capable of finding the way back to their nests from great distances and we have seen on page 132 how the Manx shearwater seems to manage. Incidentally, the shearwaters were first carried 300 miles inland to Cambridge and were back in their burrows six hours later. They were also successful when released from Venice, a town with which they could not possibly have been familiar.

Laysan albatross, taken from Midway Island as they were a danger to aircraft, and carried 3,200 miles to a suitable site, were back ten days later. One bird even returned from 4,100 miles. White storks have found their nests after being released 1,400 miles away, while swallows and swifts have made it from 1,000 miles. Also sandmartins, bank swallows and of course pigeons are expert homers. One of the most remarkable of these homings was that of a Leach's petrel, the size of a blackbird, which covered 3,000 miles in a fortnight. One presumes the methods used were those employed by the shearwater, namely position-finding from the Sun using a sense of time.

In Hawaii, pigeons are raced at night but the islands are small, the sky is generally clear and, apart from coastlines, the birds can probably navigate visually by the glow of town lights. We also know pigeons can home towards certain natural and man-made sounds and sense magnetic dip. When moved from its loft across the lines of magnetic force, infinitesimal currents would be set up in different directions according to whether the bird was moved east or west and these might be summed at least to the point at which the bird might know which way to turn. It would then be necessary for the pigeon to fly until the magnetic dip were right, travelling along a line of magnetic force, and then turn either east or west. It is known that birds have become lost when the lines of force have split into two.

Alternatively birds can be directed by the stars and perhaps also by the Moon. Cranes certainly orient themselves by the Sun by day and they also travel at night. They have been known to be confused by the lights of towns which, to them, may seem like stars in an upside down sky and they have then flown aimlessly around until the gradual dimming of the lights has encouraged them to continue on their ways.

It has been claimed by K. Schmidt-Koenig that pigeons can find their ways to within one or two miles of their lofts even when wearing frosted glasses over their eyes, which makes the final stage of homing much more difficult.[61] The birds must therefore be using a system not depending on eyes and this rules out astronomical methods. These are also eliminated by the requirement to measure and remember a vertical to within 1/40°, far better than the 1° indicated by other tests, and the same precision would be needed if magnetic homing were postulated. It has been suggested that pigeons are able to detect the scent round their lofts, particularly when carried by the wind.[62]

Migration. Natural selection would not have favoured migration had there not been overriding advantages. Serious losses occur during the journeys and also from the need to get used to new and unfamiliar surroundings. In equatorial forests, birds can live on insects and seeds all the year round and those that inhabit tropical areas with a wet and a dry season need not travel far to where there is plenty of food.

The migrants are those that nest in the northern parts of Europe, Asia and America in the summer months, taking advantage of the long hours of daylight in which to hunt for food to bring up their nestlings. When it gets cold, seeds and insects disappear and the birds then make long journeys, often migrating across the tropics and deep into the southern hemisphere. There are no comparable migrations from the southern hemisphere simply because there are no great land masses below latitude 40° south, except for inhospitable Antarctica.

As we saw on page 38, birds brought up alone from the egg become restless at the time of year their parents migrate and flutter in the direction in which they have to travel. This shows that migration is an inborn urge. Indeed, young birds tend to migrate further than their parents, who may learn by experience that they need not travel quite so far. Also, young birds do not always fly away at the same time as their parents. Some set out afterwards and others start before.

It is certain the instinct to migrate is very strong. Garden warblers migrate from northern Europe in autumn and follow the west coast of Europe to Gibraltar and then down the west African coast. On reaching Angola, they are in ideal conditions but, after a short rest, they press on a couple of hundred miles into desert lands, where conditions are extremely harsh, and there they stay throughout the northern winter. This seems to apply to the old birds as well as the young!

It is also worth noting that, as a general rule, birds that live on seeds, such as storks, turtledoves, chaffinch, starlings and also predators including hawks and pelican, generally migrate by day but allow time off for food. However, bunting, curlew, flycatchers, orioles, warblers and most insect catchers migrate at night and may feed in the day, but the birds that live on aerial plankton in the form of tiny creatures that float in the air, namely swifts and swallows, travel by day and may feed as they fly.

It seems highly likely that young birds in particular tend to migrate by keeping to an instinctive direction and travelling for a time sensed by various means. For the journeys home, birds remember, in the most extraordinary way, the routes they took on the way out and the parents and those reared early enough the previous year will remember the country round their original nests. For, as we know, birds often return to the same barn, bush or tree from which they set out six months earlier.

This difference in behaviour between young and older birds was well illustrated by an experiment with Baltic starlings which migrate by day west-south-west through Holland to northern France, southern England and southern Ireland. Dr. A. C. Perdeck caught 11,000 of these birds in Holland, marked them and then carried them 400 miles at right angles to Switzerland.[63] The young birds took up their original courses and ended up in Spain and Portugal, but a number of the older starlings realised something was wrong and managed to find their way to northern France. What was even more significant was that, on the way back, the young birds naturally remembered the way to Switzerland but then, not having flown there and learnt the way, were unable to proceed. Similar displacements of the release points of hooded crows have produced parallel results.

Young golden plovers even migrate along routes that differ from those used by their parents. Nesting in the far north of Canada and even in Alaska, the old birds travel first to Labrador and then turn southwards to fly 2,500 miles across the Atlantic to Brazil, continuing on to the Argentine pampas, 8,000 miles from where they started. The young birds, who migrate from their nests after the parents have left, travel overland via the central United States and Panama, a route which, oddly enough, both old and young birds follow on the way home. As opposed to this, some birds such as geese and ducks travel in flocks, the young with the old, led apparently by different birds each year but presumably navigating almost entirely by memory. Thus the greylag goose uses traditional routes and standard resting places.

Incidentally, not all the golden plovers from Alaska go to the Argentine. Some of these land-birds, with their relatives in Siberia, travel 2,000 miles across the open sea to the Hawaiian Islands, which fortunately stretch along a line also about 2,000 miles in length across their path. These birds may eventually reach Australia and New Zealand either by that route or via the west coast of Asia, and the bar-tailed godwit makes a similar journey, the birds returning to their northern grounds in early summer. For most of these migrations, the easterly winds in the tropics shown on page 147, will keep the birds along the coast of Asia and the East Indies but, in northern waters, the westerlies will drift them towards Canada.

Many migrations are affected by topography. In North America, for example, birds tend to follow one of four 'flyways' illustrated in Fig. 41. These run down the sides of the Rocky Mountains and the Appalachians. Many birds like to fly low when they migrate and they may keep to sheltered valleys and follow rivers. Chaffinch and meadow pipits may alter course by as much as 90° to follow such a path. On coming to a coastline at an angle, they tend to drift down it until they can see a way across or, if they do venture out to sea, they turn back if no land is visible ahead and continue to travel down the coast.

Storks and birds of prey rely on soaring flight and follow upcurrents of air, particularly those over the slopes of mountains. Indeed, they may not get airborne until the Sun has warmed the ground sufficiently for them to be able to take advantage of upward air movements. Naturally, these birds dislike flying over the sea where there are few vertical currents. So, when migrating from Europe to Africa, they avoid the Mediterranean. If they start in western Europe, they fly over Gibraltar on their way to the Gulf of Guinea or, if from further east, they cross over by the Bosphorus and follow the Nile down to South Africa. Similarly, cuckoos play it safe by crossing the Mediterranean via Sardinia and swifts shorten their sea crossing by flying through Greece along auxiliary routes shown in Fig. 41.

The European swallows, swifts and cuckoos fly down to South Africa to winter and birds from Siberia make similar pilgrimages to avoid the barrier of the Himalayas, while others to the east make their way down the China coast to Malaysia and often on to Australia. Yet some migrants seem deliberately to ignore topography and will travel through high passes at 20,000 feet, even flying at heights which a small detour would avoid. Tiny humming birds, smaller than a man's little finger, are found as far north as Alaska, and have crossed the

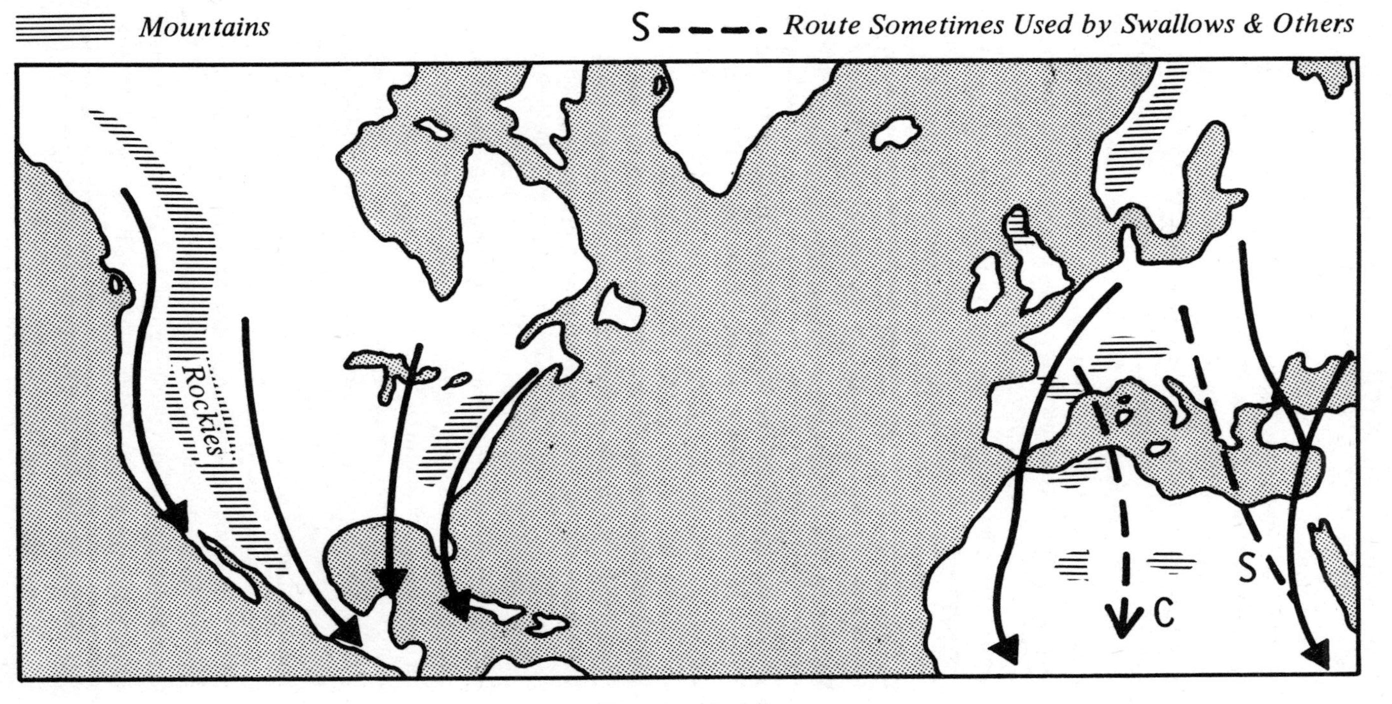

41. Topographical flyways

Rocky Mountains at 12,000 feet, and their southern cousins regularly fly 500 miles across the Gulf of Mexico in their search for suitable flowers and insects.

Although there are fewer land masses in temperate regions in the southern hemisphere, there are some interesting land-bird migrations. Long-tailed cuckoos leave New Zealand in March, the end of the southern summer, and fly over a huge ocean area stretching north-west and north-east as far as the Carolines, Marshall Islands and Pitcairns but concentrating on the area round Samoa and Tonga. There are also some oddities. A bee-eater from Madagascar struggles across the open ocean to East Africa to fill the gap left by another bee-eater which has just departed for the Caspian Sea and northern India.

Many people will have noticed how those high-speed flyers, the swifts and the swallows, depart in autumn as regularly as clockwork and return in the late spring equally punctually and irrespective of the weather. The return of the swallows to Capistrano, California, on or around the same day every year[64] is a well known example. Other birds, including willow-warblers, Canadian geese and American robins, seem to be influenced by temperature and will return early if the weather is favourable, presumably by taking off into a warm wind which promises better conditions and more food and insects. Indeed many birds, particularly small ones,[65] wait for winds coming from particular directions before migrating. Curlews are even said to be affected by the amount of ozone in the air, for the return of these migrants has been correlated with thundery weather.

A few types of land-bird also stage emigrations at intervals of several years but not on the same scale as insects. The waxwing lives on berries in the northern forests of Europe, Asia and America and every few years may irrupt in large flocks far to the south. Other birds also emigrate, including nutcrackers which live on seeds and nuts in the pine forests of the same area and also travel southwards, and crossbills are also believed to react similarly to shortages of food.

Birds that hardly fly at all and those which are purely landborne seem to become so attached to their terrain they do not normally migrate. However, coots have been known to march across Oregon in thousands while certain geese, ducks and grebes travel considerable distances by swimming. Of course, the best swimmers are the penguins, with flippers instead of wings. These comical little birds have been seen in large flotillas six hundred miles from the nearest land. They also go for walks on the ice, and set out north-north-east if released on the antarctic ice-cap, even when carried close to the pole. They waddle off confidently provided they can see the Sun, and this course generally leads them to the east of their rookeries on the coast of Antarctica, but the westward flowing current will then carry them back to their relatives.

The masochistic emperor penguin actually produces its young on the sea ice in the middle of the dreadful antarctic winter and takes turns with his wife to sit on the young bird when it emerges from the single egg. When the spring comes, and the appalling weather gradually eases, the birds and their single chicks travel away from the pole into less rigorous conditions. Many happily embark on a summer cruise on a convenient raft of ice.

Sea-birds. Surely the most striking of all migrations are those undertaken by sea-birds, which seem to travel along constant courses but to be deflected by winds and to follow, where convenient, the general directions of coastlines. Every year, the arctic tern, a little bird the size of a thrush with a rather fluttering flight, travels 25,000 miles from the Arctic to the Antarctic and back again. Thus for eight months in the year, the bird enjoys twenty-four hours of daylight.

Fig. 42 shows, on the right-hand side, the approximate routes followed. In general terms, the birds from Alaska and northern Canada travel south-south-east and those from Scandinavia and Russia go south-south-west. As a result, they follow the directions of the American, European and African coastlines but with certain exceptions. Westerly winds in the North Atlantic carry Canadian-born birds across to Africa and these then continue to the Cape of Good Hope whence they may be deflected into the Indian Ocean by the westerly winds. The north-east trade winds similarly drift the European birds over to South America.

On approaching Antarctica, the birds will be moved westwards by the polar winds. These will also help to offset drifts due to the strong westerlies in the early stages of their return journeys in which they seem to fly reciprocal courses, following remembered coastlines but crossing the Atlantic where these go in the wrong direction and become unfamiliar. Thus the arctic terns fly back to their original arctic areas.

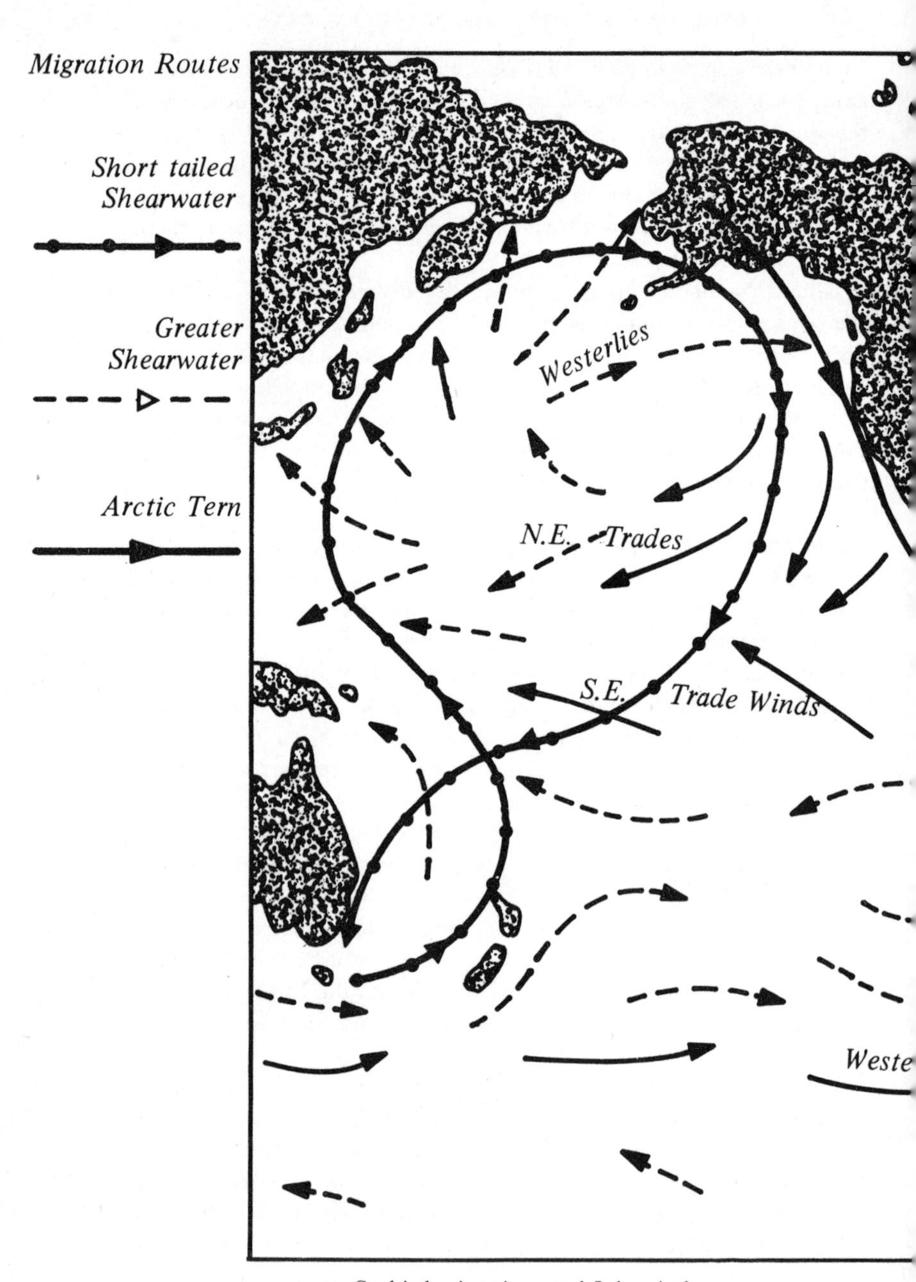

42. Seabird migrations and July winds

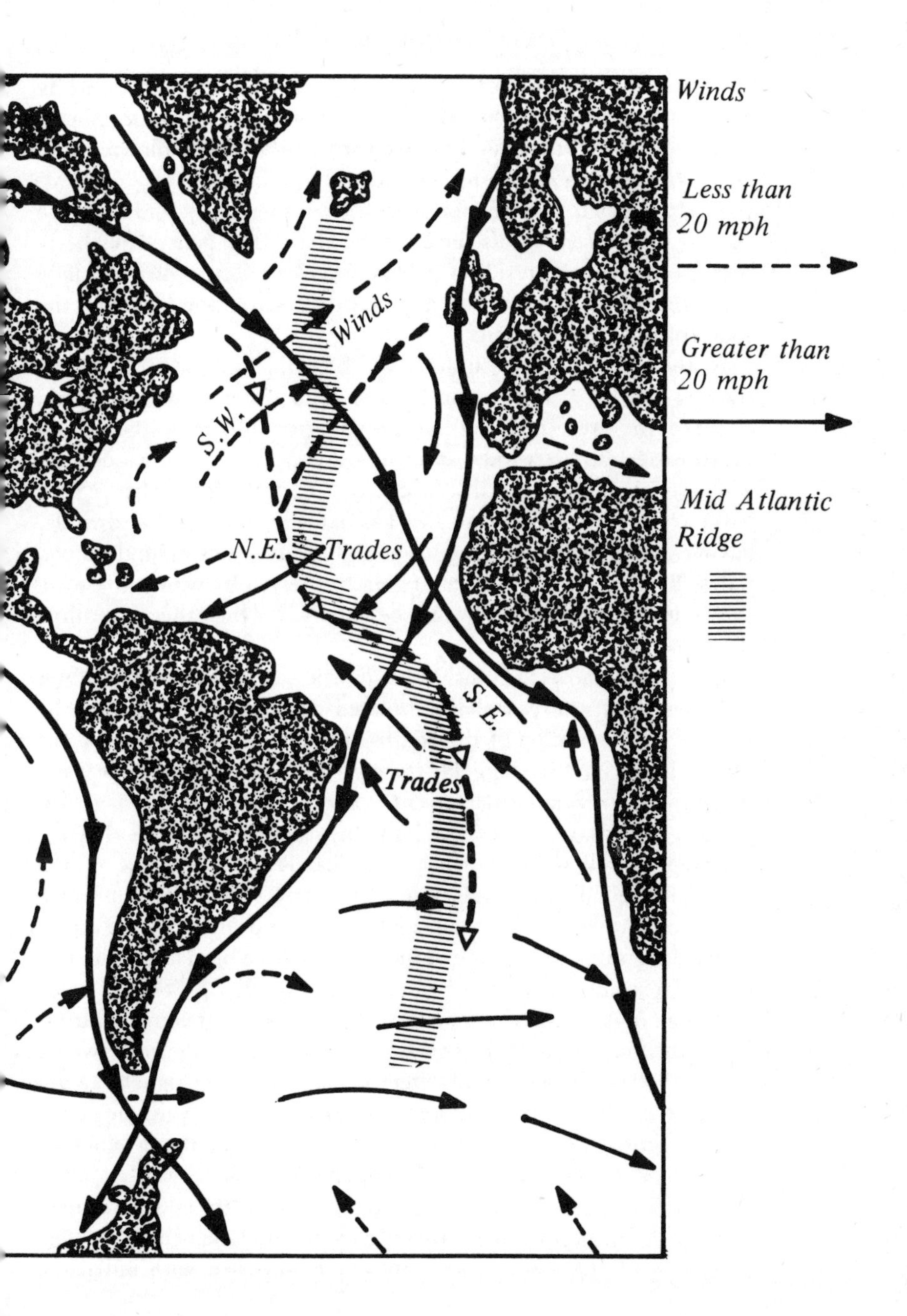
Winds
Less than
20 mph
Greater than
20 mph
Mid Atlantic
Ridge
Winds
S.W.
N.E.
Trades
S.E.
Trades

On the left-hand side of Fig. 42 can be seen the route travelled by the short-tailed shearwaters. Starting out from the area around Tasmania and flying roughly northwards, they will be blown out to sea by the very strong westerlies and will pass close to New Zealand, soon after which they will pick up the south-easterly and then the north-easterly trade winds and be carried from their general northerly route onto the western side of the Pacific Ocean. They will eventually reach the northern westerlies and will drift across the Pacific to the south of the Bering Straits.

By then, the northern summer will be dying and the birds will be starting to work their ways southwards. The westerlies will take them close to the west coast of North America and then the strong north-easterly trade winds, and the south-easterly trade winds also, will carry them to the north-east coast of Australia from whence they will have to struggle back to the Tasmania area against headwinds, having completed a huge figure-of-eight journey of around 20,000 miles. Because this journey is so greatly affected by winds, these are shown on the map as they would be in July, for then they determine the route almost entirely.

Fig. 42 also shows the southwards journey of the greater shearwater which nests in the completely isolated island group of Tristan da Cunha, in the middle of the south Atlantic. These birds leave the islands in spring, travel quickly through the tropics where there is not much food, and roam widely over the north Atlantic. For their return journey, they have to hit these four tiny islands. The route they are believed to follow coincides approximately with the line of the mid-Atlantic ridge, an underwater mountain range between North and South America on one side and Europe and Africa on the other.

The land masses each side of the Atlantic have been likened to two huge plates, floating on the molten core of the Earth. This volcanic ridge has built up between them and has some interesting features. First, there are a number of transverse splits across the ridge where the magnetism changes, producing what are known as magnetic anomalies. Secondly, the ridge tends to be warmed by the volcanic action below. Thirdly, the flow of water deep in the Atlantic is generally westerly and this is particularly noticeable in the South Atlantic.

We now see that our isolated islands in fact lie at the end of a strip of water of depth comparable to that of a continental shelf and there is an upwelling of cold water, probably amply supplied with nutrients, from the deeps to the west. It may be the ridge can be detected magnetically or the water travelling over it sensed by its motion, its temperature or by certain chemical characteristics which a bird might smell or which might change the colour of the sea. All this is speculation but, in any event, the group will be identified by cloud building up over the 7,000-foot mountain on the island of Tristan long before the high ground itself appears over the horizon and there will also be the pungent scent, if one may call it that, of guano, carried downwind for long distances.

In addition to the greater shearwater, many other birds nest on Tristan da Cunha. The wandering albatross with its wingspan of twelve feet, probably circles the Earth following the westerlies, and may do the journey several times a year. Presumably, it could find the islands by leaving a known point in South America and flying east or it too might use the mid-Atlantic ridge. On one occasion, a little sandpiper from Siberia landed on this remote archipelago ten thousand miles away, but this surely must have been by accident.

Sandpipers are one of the small birds that seem fated to be blown about by winds. Every now and then, one of these birds from North America is blown across to Europe by a westerly gale and to suggest it must have travelled by ship is hardly tenable, since European land-birds are virtually unknown in Canada. Further south, tropical storms have carried European herons to the West Indies. But these are exceptions. What is so remarkable is the way birds over the ages have learnt to navigate to places which, as the world has changed its shape, have sometimes become far more difficult to find. And they have surely managed this not by orderly systems for finding the way but by using special and particular pieces of information.

MAMMALS

They evolved from reptiles but could hardly be more different. Because the young are fed from their mothers, they are taught, by virtue of this intimacy, how to search for their food and how to look after themselves. The ability to learn is of extreme value to an animal. It provides an immediate response to a change in circumstances, without the need for a genetic imprint into the brains of its distant descendants.

Mammals that live in cold climates, the extreme example perhaps being the polar bear, grow thick fur, whereas the elephant, living in the tropics, has hardly any hairs. Even more significant is the ability of mammals to maintain their internal temperature largely irrespective of their surroundings. They may become torpid if it gets too cold and bats, hedgehogs and certain small rodents hibernate. Others that live on vegetation that dies in winter may descend from high ground into valleys, or even travel short north–south journeys towards warmer climates.

Alternatively, mammals may need to travel to another part of their preferred habitat due, for example, to a shortage of local food, and so they may travel east or west. In any event, they will generally be followed by the animals that prey upon them, which are usually mammals but sometimes birds. Although having legs which emerge from the undersides of their bodies, mammals may not be able to travel extreme distances, particularly if the terrain is unfavourable. Hence the most striking journeys undertaken are by those mammals that swim or fly.

Migrants that swim. Seals and walrus are not fully aquatic for they breed on land and, for that purpose, they may migrate, often covering six thousand miles a year. The northern fur-seal female spends the winters in the sea round Japan or California. The males, which commonly weigh a quarter of a ton, may stay round the Aleutians but set out in mid-April for the Pribilof Islands. As shown in Fig. 43, this is an isolated group to the north of the Aleutian Islands, towards which the Alaska Current flows through the first major gap in the Aleutian chain. It also happens that the Earth's magnetic field leads through this same gap to the islands.

By May, the male fur-seals have driven the local seals off the islands and the battles for territory have begun. For weeks on end, no bull dare leave his territory even to feed until, in mid-June, the females arrive. Each produces a single pup and then mates for the next year. Meanwhile, the younger males and females who have followed their mothers are forced to look on from a respectful distance.

By August, the bulls depart but stay for a time in northern waters and, at about this time, the pups from the previous year arrive. It begins to get cold in October and so the seals travel back to Japan and California. Other seals follow a similar north-and-south pattern, some

43. Summer migrations of seals

Atlantic seals breeding far to the north and even using drifting icebergs instead of beaches. Seals in the southern hemisphere similarly undertake south and then north migrations.

It is amazing that fur-seal pups, which travel to the Pribilof Islands after their mothers, can have learnt the way on the journey back the previous year but perhaps they have an innate memory of the route. In any event, a female harp-seal drops her pups onto an icefield in the spring either off Newfoundland or Jan Mayen, an island east of Greenland, and these fields drift southwards and break up, yet the pups return to their nurseries. As we have seen, seals also have a crude form of sonar but this is unlikely to help them to find their way. However, it is likely they are able to orient themselves magnetically, because they must be very concerned about their directions when travelling under sea-ice. Indeed, when fishing from blow-holes, they are believed to travel up and down vertically to avoid losing the way back to their only source of air. Also, unlike the completely aquatic mammals, they retain a sense of smell though its effectiveness underwater may be minimal.

Whales and dolphin live in a world of fishes, coming up to the surface only to breathe. They have replaced the lateral line by hearing and by sonar and also have very delicate tactile sensors all leading into large brain lobes. The biggest baleen whales, the largest animals ever to have lived on our planet, weigh as much as 1,600 men and feed on crustaceans an inch or so long which, apart from the north-west coast of Africa, the Bay of Bengal and the Gulf of Aden, are found mainly in the cold waters around Antarctica. In the southern winter, the ice spreads over the sea and the whales migrate to their breeding grounds close to the tropics and not far from the great continents, returning to the Antarctic when the ice recedes in the early southern summer.

There are also baleen whales in the Northern hemisphere. In winter, the grey whale travels from the Bering Sea or the coast of Siberia southwards to California to breed, the young not having the necessary blubber to survive an arctic winter. The families move back northwards in April, probably following the coast and using their low-frequency sonar to identify rocks and shallows.[66] However, most of the northern whales are of the smaller toothed varieties and feed on squid and herrings as well as crustaceans.

The toothed whales with their very precise high-frequency sonar are able to recognise small objects close by. They feed in arctic waters along the northern coasts of Russia and Canada, and migrate down the coasts of America and Asia, even entering the Indian Ocean via the Pacific or the Atlantic. Whales are known to follow fixed routes, known to the whalers of old as the 'rich veins', and these are probably not chosen for their scent as whales have failed to readapt their olfactory systems to be usable in the sea.

Whales are highly intelligent creatures and can be expected to have very good memories and probably navigate in the oceans much as birds do in the air, by remembering ocean currents, detecting water temperatures and recollecting topographical features underwater, using their very sensitive sonars. The smaller members of the whale family, the dolphins, are distributed widely in temperate waters but porpoises are fond of coastlines and estuaries and have even been caught far up the Rhine at Cologne.

Bats. Turning now to mammals that fly, we shall exclude those that merely glide and discuss only the true fliers, the bats. They may weave about uncertainly when hunting in order to deceive their prey, but they can fly fast and have even been seen in company with swallows, though they prefer to travel below 200 feet. Bats hibernate and, when they do, become cold-blooded. They may wake up to travel in flocks for a few tens of miles in order to hibernate in a cave or building which they prefer to their present shelter. Yet many make longer journeys. They travel by night from one known stopping place to another, but continue during the day if they fail to make their objective. They will even cross mountains, following older bats and in spite of heavy losses.

Several species of bat in the United States spend the summer either side of the Canadian border but winter in the southern states. Frequently the bats follow the coast, flying at some distance out to sea, sometimes as far as fifteen miles, and they may even reach Bermuda. The rufous bat follows this pattern but spends his summer in the far north of Canada. The millions of bats which inhabit the Texan caves have already been mentioned. Most of the females travel to Mexico in the autumn, leaving the less venturesome males behind.

Generally, European and Asian bats do not travel so far. In Russia, small bats travel short distances to hibernate. Bats from North-east and North-west Germany fly a couple of hundred miles from their summer quarters to hibernate in central Germany, returning to their summer quarters in March or April. The mouse-eared bat is however an exception to the general rule. The males live in caves in the Atlas Mountains and, at the end of August, the females fly in, having travelled from central Europe over the Pyrenees, where they have resting caves, across Spain, more stop-over caves in Gibraltar, and along Morocco (the caves of Hercules) and so to their hibernation caves. They eat to store up food, mate and then go to sleep for the winter.

In April and May, the females return to their summer quarters living in little colonies. The young are generally born on the same day in each colony and are fed for a month, after which they are taught to hunt. The summer ends, autumn begins and the bats are seized by migration restlessness, squeaking and flying in all directions and, as soon as it is dark, they set off once more for the Atlas Mountains in Africa.

It is difficult to judge how much use bats make of their sonar to find their ways for, if deprived temporarily of its use, they refuse to fly. This is natural for presumably they rely on it to avoid hitting unknown obstacles, but they can manage if they are blindfolded and seem to be able to home over short distances by some form of orientation. Bats taken a hundred miles away to a place unknown to them have been known to find their ways home. What was more extraordinary, when carried by train in closed boxes, the bats became most excited and squeaked madly when the railway line passed close by their caves. Cynics would suggest they could smell the odour.

The non-insect-eating fruit-bats or flying foxes make regular migrations following the ripening of crops in tropical regions and particularly in Australia, where they travel from Queensland and may reach Victoria by mid-summer. They generally fly by day, though the tailed fruit-bat migrates by night, avoiding obstacles by clicking the tongue to produce a crude form of sonar. Tomb-bats, which roost in places where it is pitch dark, also produce sonar noises by slapping the tongue against the roof of the mouth, but they generally fly about by day, using their eyes.

Land migrants. The spread of human beings over the land, setting aside areas for agriculture and developing communications such as roads and railways which tend to act as barriers across the countryside, all these have inhibited the travels of land mammals, though tunnels are built below motorways to allow small animals to pass from one side to the other. Nevertheless, railways in the United States have followed the trails of the great herds of bison that once migrated north and south across the prairies along regular routes, a process of natural selection having developed these routes along easy gradients.

In Africa, the trails of elephant herds, which retreat into the shade of forests in the hot season but travel long distances out into open spaces when the rains come, show how constant are the routes taken by migrating land mammals. Even today, large numbers of antelope, buffalo, kudu, gnu, zebra and wildebeest may journey up to a thousand miles from one feeding ground to another along traditional paths. The springboks, able to jump eight feet into the air and land at the same spot, used to migrate in huge numbers to high ground in South-west Africa to eat the growing grass. Later, they would travel back north-east to land freshened by the rains, though they would delay this journey if the rains had failed, which suggests that atmospheric conditions influenced their return migrations.

In the far north of Europe, Asia and America, man has not penetrated sufficiently to prevent migration. In Canada, herds of caribou, the name given to American reindeer, manage to eke out an existence in the barren wastelands around the Arctic Circle during late spring and early summer. In July, they migrate southwards into the forests, following well defined routes, where they find little undergrowth but plenty of suitable mosses. They stay there until late February or early March before moving northwards again, travelling ten miles or so each day in large herds, swimming across rivers and walking delicately in single file across frozen lakes. In June, the females wander away to bear their calves. In all these migrations, the caribou are accompanied by predators, particularly packs of wolves.

In Europe and Asia, the reindeer undertake similar migrations led apparently by large females. In Scandinavia and Lapland, the reindeer live during the summer on the rich grass in the mountains, free from the pestilent mosquitos. In autumn, guided by the local inhabitants, they go down to lower land and, in the spring, travel to the coast. The Laplanders are again on hand to help young animals on their journeys to offshore islands and, in return, they take their toll of the less vigorous members of the herds.

In eastern Siberia, large numbers of reindeer spend their summers in the mountains and descend, like the caribou, to the forests in the winter, going back to the mountains in the spring partly to avoid the huge swarms of mosquitos hatched among the trees. In central Siberia, the saiga-antelope, a thick-legged animal with a nose like a tapir's, moves north and south over the steppes as the snow line retreats and advances and, by this means, covers huge distances. Similar but shorter journeys are made by elk and moose.

The most celebrated emigrant among the mammals has been that attractive little six-inch long rodent, the lemming, which generally lives quietly in a burrow in summer and in a nest made of leaves in the winter. The females produce four or five offspring a year, sufficient to offset the depredations of dogs, cats, ermines, foxes, lynx, owls, seagulls, skuas, weasels and other Scandinavian predators. Then, quite suddenly, after a period when plenty of seeds and insects have been available, the female has started to produce bigger litters not once but four or five times a year. Instead of a couple of her offspring surviving, there might well be fifty.

Like the locusts, the lemmings would react to this gross overcrowding of their burrows by staging a mass exodus, starting with ones and twos but rapidly growing into a huge flood of little migratory rodents. Nothing was allowed to stand in their way. They would swim rivers and lakes, swarm through villages, entering houses and, though normally shy and retiring, would attack ferociously human beings in their way. Finally, arriving at the sea, they would leap into the waves and swim until they drowned in their hundreds of thousands.

Perhaps the food has not been so plentiful in recent years or the predators have become more efficient, for the lemming disasters that used to occur regularly every five years or so have gradually died away. In Canada, the lynx and the snowy owl multiply greatly in years of lemming population growth and, with the other predators, they now seem able to maintain a relatively stable situation. However, other rodents, rats, mice and snow-shoe hares may produce lemming-like emigrations when their living conditions become overcrowded, though the building of poor nests and inadequate feeding generally reduces the survival rate to one in twenty or thirty of those born.

The arguments put forward for locust emigrations, that such mass suicides are but extreme examples of a reasonable instinct to move when overcrowding occurs and that overbreeding allows for losses on the journey, may also apply to mammal emigrations. In any event, lemmings have survived as a species and therefore the disasters seem to have achieved a purpose that permits them to be classed as navigations.

Methods. For mammals, finding the way home generally seems to be a matter of retracing the route they have taken on the way out according to what they smelt, saw or felt from the topography or texture of their surroundings. Horses in particular have extremely accurate powers of recollecting where they have been. Cats and dogs, if carried away blindfold or in a box from which they cannot see out, appear able to memorise turns made over short distances and to guess at the ground traversed in between. Human beings, particularly children, may also do well at distances of a very few miles.

Many mammals mark trees and other objects with their scent and often also anoint their homes or even their families, but these actions are directed towards delineating territory or conjugal rights rather than finding the way. However, some apply scent to the feet so that their trail will be blazed and there will be no problem in finding the way home. In any event, small rodents in particular travel along runways, chosen for their cover and the absence of objects such as dry leaves and twigs which could betray their position by sound. The following of an established path is carried to extreme lengths by some animals and the elephant-shrew has already been quoted on page 25 as an outstanding example.

Small mammals seem to react to magnetism, which may help to guide them though, when the Sun is out, there will be shadows and rays of sunlight. However, it is likely that much use is made of distant objects to maintain direction, such as a line of hills or a clump of trees on the horizon. Even human beings who claim to possess a 'bump of locality' may run into difficulties in Holland where the land is flat.

Summarising, mammals rely heavily on memory to find the way. It is true that meadow voles, wood-mice and brown bats seem to have an ability to home in the same way as do edible snails. But generally, to carry a mammal away blindfold or in a box from which it cannot see out will be an unfair test. All it can do will be to wander about until it happens upon something familiar either in terms of appearance, sound or scent, which will be a matter of pure chance. Nevertheless, this same mammal will be able to navigate itself perfectly effectively around its habitat which is all that it needs to be able to do.

Apes and monkeys. Apart from certain species, such as baboons, apes and monkeys live in trees and, as food and shelter is generally available all the year round, they do not need to migrate. To travel into the high branches of trees where food is normally closest to hand, the animals have developed all four legs for holding on and often tails as well and, as explained on page 82, their eyes have moved forwards in their heads to give wide binocular vision, for they cannot afford to miss a handhold.

Arboreal apes and monkeys, not being so interested in the ground, have developed the lobes of the brain that deal with sight, sound and touch rather than those which analyse smell. In particular, their feet and hands are extremely sensitive and they can appreciate the way things feel and their exact texture. Their brains, no longer dominated by smell, but involved with a wide range of subtleties requiring extremely complex cross connections, have developed enormously to the point at which it is said that a chimpanzee, arguably the most intelligent of all animals, may read a rudimentary map,[67] showing the emergence of spatial imagination.

An animal that lives in trees cannot use distant objects to orient itself and often may not be able to see the Sun clearly and will find it difficult to recognise stars. It must therefore rely on its memory. The behaviour of a lemur in a large cage with plenty of climbing area illustrates this very nicely. The animal will start by exploring its territory carefully, testing the branches and ledges with great caution to make sure they will bear the weight and that they are not slippery. Only then will it start to travel round, first slowly and then at ever increasing speed, yet always putting hand or foot in exactly the same place on successive branches or footholds.

One would not assess the ability of a monkey to navigate by whether it can find its way home from a place it has never been to. In real life, it would not be so stupid as to travel without noticing where it had gone. Yet Dr. R. R. Baker has carried students away from their university by winding routes. The transport was stopped and the students were able to point towards the university with an accuracy which seemed too consistent to be attributed to chance. Tests with magnetic helmets showed predictable distortions of direction and so it may be that human beings, and apes and monkeys also, can make use of magnetic information.

SUMMARY

The following are selections from the text:

1. *Birds*. Rely on memory of things seen and smelt.
   (a) *Land birds*
      (i) *Fledglings* either travel with parents or use instinctive course and distance inputs.
      (ii) *Experienced birds*. Retain fine details of past journeys.
   (b) *Sea birds*
      (i) *Fledglings* travel alone using Sun and innate memories.
      (ii) *Experienced birds* are guided by Sun, coastlines and subtle signs and may also use sonar for final homing.
2. *Mammals* rely heavily on associating past journeys with things sensed. They teach their young and, except for seals, travel with them. Also
   (a) *Seals* may use magnetism and young may travel by instinctive imprints.
   (b) *Whales* may use sonar.
   (c) *Bats* use high-frequency sonar particularly for homing.

Conclusions

Stabilizers. It does not seem to have been appreciated by animal specialists how important is the part played in navigation by the stabilizers and, in particular, by their ability to sense the vertical. There may be two reasons for this. First, in human beings, the sense is instinctive and so it is taken for granted. Secondly, animal experts have naturally sought advice from mariners, on whose past efforts the whole of modern navigation has been developed. But, because ships pitch and roll, mariners have traditionally used the sea horizon instead of the vertical for position finding.

Specialisation. The modern mariner and airman is concerned with world-wide systems able to guide him to anywhere he wants to go and so, in navigational terms, he is a generalist. An animal, on the other hand, is a severe specialist. It only has to make its way to a specific destination and it has no interest outside those particular crumbs of information that will guide it on its individual way. It may well be that human navigators ought to study more closely the details of the routes they plan to follow.

Back-up systems. In recent years, aircraft have begun to carry not one inertial navigation system but three so that the odd man out may be ignored. Animals duplicate or triplicate their information by combining inputs from several senses which has the great advantage that, if by some circumstance such as weather one sense is unavailable, the others can take its place, at least on a temporary basis. Thus, as we have seen, a pigeon steering by magnetism and by the Sun will fly quite happily over areas of magnetic anomalies.

Summary. We may summarise the three major points, *specialisation*, *stabilizers* and *back-up systems* by using a human example. Through the islands where a famous Polynesian navigator, Kaho Mo Vailaha, once lived,[68] ocean swells rolled steadily from the east winds to the north and the west winds in the south. As each swell approached an island, its face would curve and it would be reflected. When it had passed the land, the swells would join up along a line of turbulence. This was the *special* situation.

Lying on his back in his catamaran, Kaho would sense these swells and, using his *stabilizers*, separate one from the other and distinguish the minute distortions due to the various islands. In this way, he would pick his way with certainty through unseen atolls. For confirmation, he would dip his hand in the water and could sometimes tell where he was merely by its temperature and taste, for he had lost his main *back-up system*. Kaho was blind.

References

1. Anderson E. W. (1979), *J. Inst. Navigation*, Vol 33, No. 1.
2. *Shorter Oxford English Dictionary* (1950), Oxford University Press, London.
3. Shamburgh G. E. (1974), *Encyclopaedia Britannica*, Vol 5 (p. 1132) Encyclopaedia Britannica Inc., Chicago.
4. Dijkgraffe S. (1974), *Encyclopaedia Britannica*, Vol 11 (p. 807) Encyclopaedia Britannica Inc., Chicago.
5. Gatty H. (1969), *Nature Is Your Guide*, Collins, London.
6. Droscher V. B. (1969), *The Magic of the Senses*, Allen, London.
7. Burton M. (1973), *The Sixth Sense of Animals*, Dent, London.
8. Pittendrich C. S. (1959), Photoperiodism and related phenomena in plants and animals, *AAA*, Washington DC.
9. Brown F. A. (1974), *Encyclopaedia Britannica*, Vol 14 (p. 69) Encyclopaedia Britannica Inc., Chicago.
10. Burton M. (1973), *The Sixth Sense of Animals*, Dent, London.
11. Brown F. A. (1954), *American Journal of Physiology*, No. 178.
12. Clarke L. R. (1939), Observations on the palolo, *Carnegie Institution of Washington year book*, No. 73.
13. Geisler M. (1961), *Zeitschrift für Tierpsychologie*, No. 18.
14. Schmidt-Koenig K. (1975), *Migration and Homing in Animals*, Springer-Verlag, Berlin.
15. Matthews G. V. T. (1968), *Bird Navigation*, Cambridge University Press.
16. Droscher V. B. (1969), *The Magic of the Senses*, Allen, London.
17. Frisch K. von (1968), *Dance Language and Orientation of Bees*, Harvard University Press.
18. Kramer G. (1957), Experiments on bird orientation and their interpretation, *Journal of Ornithology,* No. 99.
19. Wiltschko W. and R. (1972), The magnetic compass of European robins, *Science*, No. 176.
20. Baker R. R. (1980), *New Scientist*, 18 Sep.
21. Leask M. J. M. (1976), *Nature*, Vol 267, London.
22. Schmidt-Koenig K. (1979), *Avian Orientation and Navigation*, Academic Press, London.
23. MacNichol E. F. Jnr (1964), Three-pigment colour vision, *Scientific American*, Vol 211, No. 6.
24. Hajos A. (1964), Optical errors of eyes, *Umschau*, Vol 64.
25. Fender D. H. (1964), Control mechanisms of the eye, *Scientific American*, Vol 211, No. 1.

26. Lettvin J. Y. (1961), Remarks on the visual system of the frog, *Sensory Communications*, Massachusetts Institute of Technology Press.
27. Muntz W. R. A. (1962), What the frog's eye tells the frog's brain, *Journal of Neurophysiology*, Nov. 1962.
28. McElroy W. D. and Seliger H. H. (1962), Biological luminescence, *Scientific American*, Vol 207, No. 6.
29. Beebe W. (1951), *Half-mile down*, Duell Sloan and Pearce, New York.
30. Bullock T. H. and Cowles R. B. (1952), Physiology of an infra-red receptor—the facial pits of vipers, *Science*, Vol 115.
31. Burton M. (1970), *Animal Senses*, David and Charles, Newton Abbot, UK.
32. Rosenweig M. R. (1961), Auditory location, *Scientific American*, Vol 205, No. 4.
33. Sparks J. and Soper T. (1970), *Owls*, David and Charles, Newton Abbot, UK.
34. Griffin D. R. (1958), *Listening in the Dark*, Yale University Press.
35. Roeder K. D. and Asher E. (1961), The detection and evasion of bats by moths, *American Scientist*, Vol 9, No. 2.
36. Dunning D. C. (1968), Moths are warning bats, *Zeitschrift für Tierpsychologie*, Vol 25, No. 2.
37. Lineham E. J. (1972), *National Geographic*, Vol 155, No. 4.
38. Smith S. L. (1961), Clam digging behaviour in starfish, *Behaviour*, Vol 18.
39. Kascher A. H. (1959), On the behaviour of Lariophagus distinguendus (chalcis beetle), *Behaviour*, Vol 14.
40. Heatwole H., Davis D. M. and Wenner A. M. (1962), The behaviour of Megarhyssa (ichneumon-fly), *Zeitschrift für Tierpsychologie*, Vol 19.
41. Droscher V. B. (1965), *The Mysterious Senses of Animals*, Hodder and Stoughton, London and Dutton, New York.
42. Burton R. (1973), *The Life and Death of Whales*, André Deutsch, London.
43. Harder W. (1965), Electrical fish, *Umschau*, Vol 65.
44. Lissmann M. W. (1960), Electric location by fishes, *Scientific American*, Vol 208, No. 3.
45. Yeagley H. L. (1951), A preliminary study of a physical basis for bird navigation, *Journal of Applied Physics*, No. 18.
46. Barlow J. S. (1963), Intertial navigation as a basis for animal navigation, *Journal of Theoretical Biology*, Vol 6.
47. Droscher V. B. (1969), *The Magic of the Senses*, Allen, London.
48. Sauer F. and E. (1960), Star navigation of nocturnal migrating birds—the 1958 planetarium experiments, *Biology*, Vol 25.
49. Clemens W. A., Forester R. E. and Pitchard A. L. (1939), Migration and conservation of salmon, *Publication of American Association for the Advancement of Science*, Vol 8.

50. Hasler A. D. (1966), *Underwater Guideposts — Homing of Salmon,* University of Wisconsin Press.
51. Frisch O. von (1959), *Animal Navigation*, Collins, London.
52. Williams C. B. (1958), *Insect Migration*, Collins, London.
53. Rainey R. C. (1963), Meteorology and the migration of locusts, *World Meteorological Organisation Technical Notes*, Vol 54.
54. Brower L. P. (1961), Studies on the migration of the monarch butterfly, *Ecology*, Vol 54.
55. Street P. (1976), *Animal Migration and Navigation*, David and Charles, Newton Abbot, UK.
56. Schmidt J. (1932), Danish eel investigations during 25 years, *Carlsberg Foundation*, Copenhagen.
57. Oshima K., Hahn W. E. and Gorbman A. (1969), Electro-encephalographic olfactory response in adult salmon to waters traversed in homing migration, *Journal of the Fish Reservation Board*, Canada, Vol 26.
58. Twitty V. (1959), Migration and speciation in newts, *Science*, Vol 130.
59. Carr A. (1972), Orientation problems in the high-seas travel and terrestrial movements of marine turtles, *Biotelemetry*, Pergamon Press, New York.
60. Dorst J. (1962), *The Navigation of Birds*, Heinemann, London.
61. Schmidt-Koenig K. (1979), *Avian Orientation and Navigation*, Academic Press, London.
62. Papi F., Ioalé L., Fiaschi V., Benvenuti S. and Baldaccini N. E. (1974), Olfactory navigation of pigeons, *Journal of Comparative Physiology*, Vol 94.
63. Perdeck A. C. (1958), Two types of orientation in migrating starlings, *Ardea*, Vol 46.
64. Brown F. A. Jnr (1977), *Encyclopaedia Americana*, Americana Corporation, New York.
65. Richardson W. J. (1978), Timing and amount of bird migration in relation to weather, *Oikos*, Vol 30.
66. Reysenbach de Haan F. W. (1966), *Listening Underwater*, University of California Press.
67. Menzel E. W., Romack D. and Woodruff G. (1978), Map reading by chimpanzees, *Folia Primat*, Vol 29.
68. Lewis D. (1972), *We the Navigators*, Australian National University Press, Canberra.

Index